TUTORIAL CHEMISTRY TEXTS

7

Organotransition Metal Chemistry

ANTHONY F. HILL

Australian National University, Canberra

RS•C

ROYAL SOCIETY OF CHEMISTRY

Cover images © Murray Robertson/visual elements 1998–99, taken from the
109 Visual Elements Periodic Table, available at www.chemsoc.org/viselements

ISBN 0-85404-622-4

A catalogue record for this book is available from the British Library

Published by The Royal Society of Chemistry, Thomas Graham House, Science Park,
Milton Road, Cambridge CB4 0WF, UK
Registered Charity No. 207890
For further information see our web site at www.rsc.org

Typeset in Great Britain by Wyvern 21, Bristol
Printed and bound by Polestar Wheatons Ltd, Exeter

Preface

Take the crowning achievements of the various sub-disciplines of chemistry, combine them and you have organotransition metal chemistry. First, the organic chemist's shrewd manipulation of functional groups, allowing strategic multi-step synthesis. Add to this the main-group chemist's understanding of reactivity within a framework of periodicity. Apply this to the colourful playground of the transition elements with their enormous variations in oxidation states, ligands, coordination numbers and associated stereochemistries. The scope for intellectual pursuit is staggering, but organometallic chemistry is not simply an academic exercise. It underpins industrial chemical processes on a megatonne scale, providing commodity chemicals from the simplest petrochemicals to the most exotic pharmaceuticals.

This text covers the material I consider appropriate for a core introductory course in organotransition metal chemistry. For all students, such a course will be their first encounter; for some it will be their last, depending on their degree specialization in later years. The material covered here should be seen as the basic tools which any graduate might be expected to call upon. It cannot do justice to the enormous breadth of the subject and the exciting tangential fields (*e.g.*, polymers, metals in catalytic and stoichiometric organic synthesis, industrial chemistry). Tutorial examples, problems and answers are available on the RSC's Tutorial Chemistry Texts website at http://www.chemsoc.org/pdf/tct/organoexamples.pdf, http://www.chemsoc.org/pdf/tct/organoproblems.pdf and http://www.chemsoc.org/pdf/tct/organoanswers.pdf.

I would like to acknowledge the support during the preparation of this text of my partner Mark, to whom it is dedicated.

Anthony F. Hill
Canberra, Australia

TUTORIAL CHEMISTRY TEXTS

EDITOR-IN-CHIEF

Professor E W Abel

EXECUTIVE EDITORS

Professor A G Davies
Professor D Phillips
Professor J D Woollins

EDUCATIONAL CONSULTANT

Mr M Berry

This series of books consists of short, single-topic or modular texts, concentrating on the fundamental areas of chemistry taught in undergraduate science courses. Each book provides a concise account of the basic principles underlying a given subject, embodying an independent-learning philosophy and including worked examples. The one topic, one book approach ensures that the series is adaptable to chemistry courses across a variety of institutions.

Further information about this series is available at www.chemsoc.orgltct

Orders and enquiries should be sent to:
Sales and Customer Care, Royal Society of Chemistry, Thomas Graham House, Science Park, Milton Road, Cambridge CB4 0WF, UK

Tel: +44 1223 432360; Fax: +44 1223 426017; Email: sales@rsc.org

Contents

1

Introduction, Scope and Bonding

Aims

By the end of this chapter you should have a feeling for:

- The range of compounds that constitute the important classes of organotransition metal ligands covered in later chapters
- The electronic book-keeping convention (18-electron rule) used to describe such compounds
- The qualitative bonding of transition metals to selected organic fragments

1.1   What is Organometallic Chemistry?

The subdivision of chemistry into inorganic and organic domains reflects history not nature. Organometallic chemistry, the subject of this text, is one unifying point of contact between these two disciplines; it embraces and enriches both. Organometallic chemistry is concerned with the metal–carbon bond in all its many and remarkably various forms. Organotransition metal chemistry (compounds featuring bonds between carbon and a transition metal) has matured in the last five decades, in parallel with our general understanding of the bonding of more classical ligands to transition metals. The laboratory curiosities and exotic compounds of yesteryear are today routinely the synthetic organic chemist's reagents, the polymer chemist's catalysts, the industrial chemist's meal-ticket. The eventual exploitation of today's organometallic curiosities provides the challenge and opportunities for this generation of readers.

We shall consider organometallic compounds as those in which the 'metal' has a comparable or lower Pauling electronegativity (PE) than that of carbon (2.5). For transition metals, these span the range 1.3

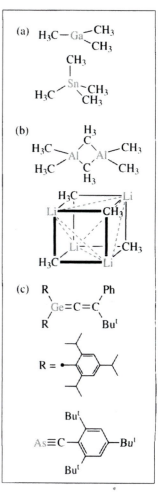

Figure 1.1 Main group M–C bonding

The discussion assumes a
familiarity with the basic concepts
of transition metal coordination
chemistry, as outlined for
example in *d- and f-Block
Chemistry* (C. J. Jones, RSC
Tutorial Chemistry Texts 4).

(hafnium) to 2.5 (gold). These generally increase across a transition period, and less generally down a group. Thus the first transition (3d) elements are more electropositive than the heavier 4d and 5d transition metals. The very electropositive lanthanide and actinide elements (PE 1.1–1.3) also have a rich organometallic chemistry, which will not be dealt with specifically here, however, except where it provides useful illustrative examples.

Main group elements have their own diverse organometallic chemistry, often overlapping with that of the transition metals. Primarily, this is concerned with simple element–carbon σ-bonds (Figure 1.1a). More complex structures arise for electron-deficient organometallic compounds, typically those of Groups 1, 2 and 13 (Figure 1.1b). Multiple bonding between carbon and the heavier p-block elements (pπ–pπ, Figure 1.2a) is less (though increasingly more!) common. For transition metals, multiple bonding (dπ–pπ, Figure 1.2b) is readily achieved and quite commonplace. The nature of a p-block organometallic follows primarily from the characteristics of the central metal: a typical p-block element has a predominant oxidation state [*e.g.* In(III), Sn(IV), Sb(III)], a less common though occasionally accessible oxidation state which differs by two units [*e.g.* In(I), Sn(II), Sb(V)] and a comparatively narrow range of coordination numbers and geometries rationalized by the octet and Gillespie–Nyholm (valence shell electron pair repulsion) rules . Although energetically accessible for hybridization and use in covalent bonding, the d-orbitals of these elements are unoccupied and multiple bonding to carbon, when it does arise, generally involves pπ–pπ orbital combinations. For heavier elements where this is less effective, kinetic stabilization *via* the use of sterically protective substituents is generally required (Figure 1.1c).

The defiance of these generalizations underpins the intrigue of transition elements. Some transition elements can offer as many as 11 different accessible formal oxidation states {*e.g.* [Ru(CO)$_4$]$^{2-}$ [Ru(–II)] and RuO$_4$ [Ru(VIII)]; [Cr(CO)$_4$]$^{4-}$ [Cr(–IV)] and CrF$_6$ [Cr(VI)]}; coordination numbers of 1 to 8 are observed, although 4, 5 and 6 remain the most commonly encountered. This breadth arises from the most significant feature of transition elements: the presence of partially filled d-orbitals which may be of suitable energy, symmetry, directionality and occupancy to enter into very effective multiple bonding (dπ–pπ). Indeed, even bonds of δ-symmetry become possible for transition metals. In contrast to p-block elements, the efficiency of this multiple bonding actually increases down a triad. This is due in part to relativistic effects which destabilize d-orbitals for heavy metals, increasing their π-basicity when occupied, a key factor in the bonding of such metals to unsaturated organic molecules which will be discussed later in this chapter. The enormous variation in these characteristic properties would at first glance

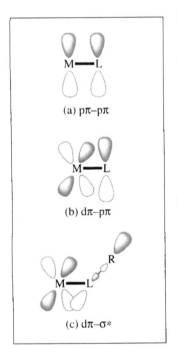

(a) pπ–pπ

(b) dπ–pπ

(c) dπ–σ*

Figure 1.2 Origins of π-bonding

make the diversity of organotransition metal chemistry appear over-whelming and even intimidating. Fortunately, various guiding principles have emerged help to conceptualize this complexity within generally well-behaved models. The most useful is the 18-electron or effective atomic number (EAN) rule, discussed below.

Two compounds serve to illustrate extremes of complexity in organotransition metal chemistry: nickel tetracarbonyl (Figure 1.3a; see also Chapter 3) and vitamin B_{12} (Figure 1.3b; see also Chapter 4). These exemplary compounds share two features: firstly, both are beautiful, the former for its simplicity, the latter for its complexity. Secondly, both are also useful, the former as an intermediate in the industrial purification of nickel, the latter as a key catalyst (metalloenzyme) in human bio-chemistry. Figure 1.3c shows a simple organocobalt compound which can serve as a model of the active site of vitamin B_{12}. Chemical models are valuable conceptual tools in understanding the chemistry of more complex chemical systems, which are themselves less amenable to direct study, when appropriate care is taken in making inferences. Many further examples of organometallic model compounds will be met in subsequent chapters.

1.2 Bonding in Organotransition Metal Compounds

It is generally accepted that d-orbitals do not play a major role in the bonding of p-block elements to carbon. The diverse chemistry of the transition metals, however, centres on the involvement of partially filled d-orbitals in the bonding to ligands. The basics of simple metal–ligand bonding are dealt with in other texts. The majority of transition metal phenomena can be accommodated by a molecular orbital or 'ligand field' treatment. In addition to providing insights into the symmetry, degeneracy and energy of d-orbitals, such a treatment also concludes that covalency is important in explaining the metal–ligand bond, even for very simple ligands such as halides, water and ammonia. For these classical ligands, however, the large electronegativity difference between the donor atom and the metal would be expected to favour largely ionic character (as is also the case for bonds between carbon and the lanthanides or actinides). For the bonding of carbon to transition metals the electronegativity difference is modest or negligible. Hence the concepts of covalency and electroneutrality become all-important, whilst the concept of oxidation state rapidly loses its usefulness. An organic chemist seldom *explicitly* considers the oxidation state of carbon. Similarly, as organometallic chemists we will only consider this (artificial) concept in the most clear-cut situations where the basic rules can be applied without ambiguity.

The important descriptors in oranotransition metal chemistry are the

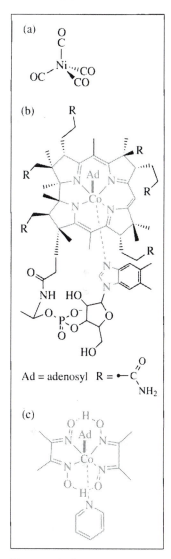

Figure 1.3 Simplicity to complexity

coordination number and the total number of valence electrons (VE), independent of where they are 'located' within the metal–ligand bonding framework (the description of which typically requires more sophisticated molecular orbital treatments). The number of valence electrons (nVE) in an organotransition metal complex is crucial to understanding the reactivity of a complex. This is described by the 18-electron rule, which is by no means absolute but provides an effective framework for interpretation.

The octet rule underpins much of the chemistry of the p-block elements, and the origins of deviations when they occur are generally well understood. An analogous generalization also arises for organometallic compounds of the d-block. The octet rule arises from the use of four valence orbitals (s + 3 p) by p-block elements in the majority of their compounds, thereby requiring eight valence electrons (8VE) to attain the effective atomic number of the next heaviest noble gas. Within the transition series it follows that nine valence orbitals are available (s + 3 p + 5 d) and that full use of these will require 18 valence electrons (18VE). Thus the 18-electron (EAN) rule emerges in its simplest form, although later we will look more deeply into the nature of this relationship. The octet rule is not absolute in p-block chemistry and the 18-electron rule has its limitations as well. Fortunately, deviations generally fall within readily understandable situations involving steric factors or more subtle electronic considerations. Note that the 18-electron rule has no useful application or predictive value in the organometallic chemistry of the f-block elements (in principle, 16 valence orbitals!). In this area, steric and electrostatic factors generally hold sway.

We shall first revisit some p-block 'octet' examples as a point of reference. The total number of valence electrons for the atom of interest in a molecule is the sum of the electrons provided by the central element in its zero-valent (oxidation state = 0) form and the ligands or substituents, with the charge of the molecule finally subtracted. Table 1.1 collates some of the ligands to be encountered later in this book, according to the number of electrons they provide. An important class of ligands which are ubiquitous in transition metal chemistry are *polyhapto* hydrocarbons and their derivatives. These are organic ligands capable of binding through two (Chapter 6) or more (Chapter 7) unsaturated carbon atoms. Each carbon through which the ligand binds provides one valence orbital and accordingly one electron to the valence count. The number of carbon (or other) atoms through which it binds (n) is referred to as the hapticity (*monohapto, dihapto, trihapto ... octahapto*) and this is given the symbol 'η^n'. By convention, if all carbons of the unsaturated system are bound to the metal, the superscript is omitted.

A selection of compounds of the simplest p-block element, boron (Figure 1.4), illustrates a number of points. The salt (1.4a) involves two

Table 1.1 Electron counting for commonly encountered ligands

nVE	Ligands
1[a]	*Ligands which in the free state would be a radical (one unpaired electron):*[a]
	H, F, Cl, Br, I, OH, OR, NH_2, NR_2, SR, PR_2, CN, N_3, NCS, bent NO, bent NNR
	Monodentate carboxylates O_2CR, dithiocarbamates S_2CNR_2, amidates $RC(NR)_2$, alkyl, aryl, vinyl (alkenyl), acetylide (alkynyl), formyl, acyl, aroyl, *etc.*
2[a]	*Ligands which in the free state would have an even number of valence electrons:*
	OH_2, NH_3, ethers, amines, thioethers, phosphines
	NR, O, S, PR
	'C_1': CO, CNR, CS, CR_2, $C=CR_2$, $C=C=CR_2$, *etc.*
	'C_2': alkenes, alkynes; molecules which bind side-on through a multiple bond: O_2, SO_2, CS_2, CSe_2, RP=PR, $R_2Si=CR_2$
3	Linear NO, nitride (N), linear NNR, NS, NSe, P
	Ligands which can be subdivided into a combination of 1VE and 2VE donations:
	e.g. η^3-allyl, η^3-cyclopropenyl, bidentate alkenyls, acyls, carboxylates, dithiocarbamates, amidates, β-diketonates, salicylates, glyoximates
4	Dienes, including cyclobutadiene and heteroatom-substituted dienes, *e.g.* vinyl ketones
5	Cyclopentadienyl, pentadienyl, tris(pyrazolyl)borate
6	Arenes, trienes (*e.g.* cycloheptatriene), thiophene, pyrrole
7	η^7-Cyclohexatrienyl ('tropylium')
8	η^8-Cyclooctatetraene
n	η^n-C_nR_x

[a]Many ligands may also carry a further pair(s) of electrons on the donor atom which may be available (if required) for π-donation, thereby alleviating an otherwise coordinatively unsaturated metal centre. 1VE ligands capable of providing a further electron pair include alkoxides, amides, nitrosyls and diazonium ions. 2VE ligands capable of providing a further two electrons include oxo, imido and alkynes

(a) $[BH_2(NH_3)_2]^+[BH_4]^-$

B (3)
+ H (2 × 1)
+ NH_3 (2 × 2)
− charge (+1)
= 8VE

B (3)
+ H (4 × 1)
− charge (−1)
= 8VE

(b) BH_3

B (3)
+ H (3 × 2)
= 6VE

(c) $BH_3(OEt_2)$

B (3)
+ H (3 × 1)
+ OEt_2 (1 × 2)
= 8VE

(d) $B(NMe_2)_3$

B (3)
+ NMe_2 (3 × 1)
= 6VE

electron-precise (8VE) boron centres that satisfy the octet rule, and the compound is stable. The molecule BH_3 (1.4b), however, has only 6VE and is therefore electron deficient. In organometallic terms, we describe this as being 'coordinatively unsaturated' with one (or more) vacant coordination site(s). The molecule cannot be isolated, but rather finds some way of relieving this unsaturation, either by dimerizing to diborane B_2H_6 or by forming adducts with Lewis bases, $H_3B–L$ (1.4c; L = diethyl ether), thereby acquiring the 2VE needed to complete the octet. The compound $B(NMe_2)_3$ (1.4d) appears to have the same electron count as BH_3 and yet is stable as a monomer. This introduces the ability of some ligands

Figure 1.4 Some boron compounds

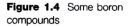

(notably those with lone pairs, *e.g.* amides NR_2, alkoxides OR, halides) to provide further electrons *via* π-donation. This phenomenon will also be encountered for such ligands in organotransition metal chemistry, in particular that of the earlier transition metals which typically have suitable vacant orbitals to accommodate π-donation from the ligands.

The examples shown in Figure 1.5 illustrate the extension of these ideas to transition metals. The first two examples each obey the 18-electron rule and are stable complexes. The manganese example (1.5c) is, however, coordinatively unsaturated (has a vacant coordination site) and cannot be isolated, but rather dimerizes *via* halide bridges in the same way that $AlCl_3$ dimerizes to $Al_2(\mu\text{-}Cl)_2Cl_4$, with each bridging halide providing three electrons to the overall count. In valence bond terms (1.5d), we can describe this situation as each halide providing a single electron to one metal (covalent bond) and an electron pair (dative bond) to the other metal. The final example (1.5e) is monomeric, although it appears to have less than 18VE. This complex provides an analogy with $B(NMe_2)_3$, in that the presence of strong π-donor oxo ligands helps to stabilize the coordinative unsaturation. Note that in both $B(NMe_2)_3$ and $Re(Me)(=O)_3$ the ligands have more electron pairs available for donation than required for 8VE or 18VE, respectively.

(a) $[Re(CO)_6]^+$	(b) $[RuCl_3(CO)_3]^-$	(c) $MnBr(CO)_4$	(d)	(e) $CH_3Re(=O)_3$
Re (7)	Ru (8)	Mn (7)	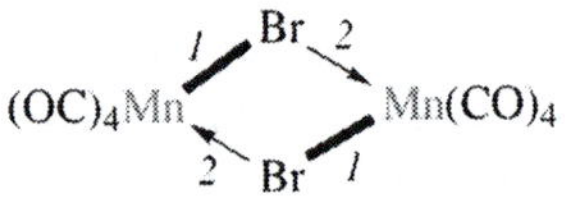	Re (7)
+ CO (6 × 2)	+ CO (3 × 2)	+ CO (4 × 2)	(OC)$_4$Mn ... Mn(CO)$_4$	+ O (3 × 2)
– charge (+1)	+ Cl (3 × 1)	+ Br (1 × 1)		+ CH$_3$ (1 × 1)
= 18VE	– charge (–1)	= 16VE	Mn (7)	= 14VE
	= 18VE		+ CO (4 × 2)	
			+ Br (1 × 1)	
			+ Br (1 × 2)	
			= 18VE	

Figure 1.5 Some transition metal compounds

The examples illustrated above involve simple ligands; however, many more complicated ligands will be encountered. In most cases we can usually subdivide complicated ligands into smaller components so long as we employ reasonable canonical (resonance) forms which each contribute the same overall number of electrons. Thus in Figure 1.6 are shown three-electron ligands, independent of the valence bond descriptions used. Figure 1.7 shows how valence electrons are counted for a range of illustrative examples, including some with metal–metal bonds. Metal–metal bonds also follow the same approach as for multiply bonded p-block compounds, *i.e.* a bond of multiplicity *n* (single, double, triple, quadruple) provides *n*VE to *each* metal.

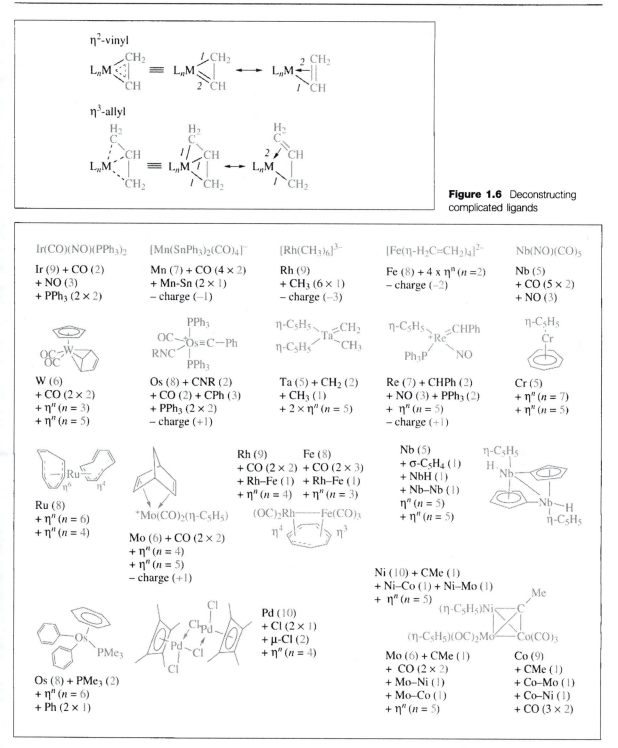

Figure 1.6 Deconstructing complicated ligands

Figure 1.7 The 18-electron rule: illustrative examples

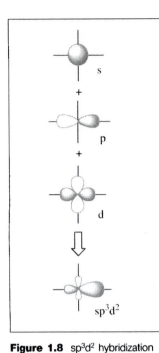

Figure 1.8 sp³d² hybridization

1.2.1 Limitations of the 18-Electron Rule

The 18-electron rule follows in its simplest form from the requirement to fully populate nine metal valence orbitals with 18VE. This is a convenient working simplification which is generally borne out by more sophisticated treatments. We can delve more deeply into the nature of the bonding, however, through a simplified molecular orbital treatment, taking for illustrative purposes an octahedral sp^3d^2 hybridized (Figure 1.8) metal centre. This perspective provides six empty sp^3d^2 orbitals, each of which is directed along *one* metal–ligand vector. This may interact with the lone pair of a ligand (σ_L) in a bonding (Figure 1.9a; M–L σ) and antibonding (Figure 1.9b; M–L σ^*) combination. The remaining three orbitals ($t_{2g} = d_{xy}, d_{xz}, d_{yz}$) protrude *between* the metal–ligand vectors and may only interact with ligand orbitals which have π-symmetry with respect to these vectors (Figure 1.9c). In the absence of ligand π-orbitals, these t_{2g} orbitals remain non-bonding in nature. Combining these interactions leads to the orbital scheme shown in Figure 1.10a. For an octahedron, therefore, the most stable configuration will require a d^6 metal configuration (6VE), in addition to the six lone pairs (12VE) (reassuringly) providing a total of 18VE.

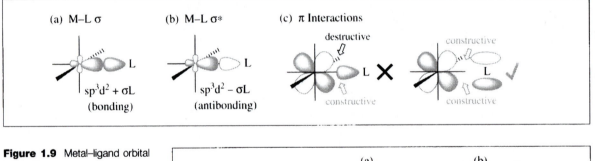

(a) M–L σ

$sp^3d^2 + \sigma L$
(bonding)

(b) M–L σ^*

$sp^3d^2 - \sigma L$
(antibonding)

(c) π Interactions

destructive

constructive

constructive

constructive

Figure 1.9 Metal–ligand orbital interactions

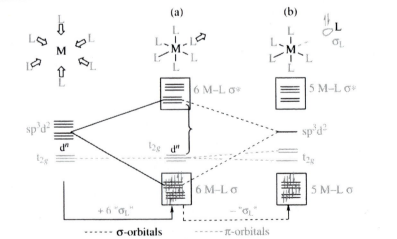

(a)

(b)

6 M–L σ^*

5 M–L σ^*

sp^3d^2

t_{2g} d^n

t_{2g} d^n

sp^3d^2

t_{2g}

6 M–L σ

5 M–L σ

+ 6 "σ_L"

"σ_L"

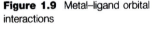

------ σ-orbitals   ------ π-orbitals

Figure 1.10 Molecular orbitals for octahedral and square-based pyramidal complexes

It is often useful to consider how this scheme would change if one ligand was then removed to provide a hypothetical 16VE complex, since this allows us to consider how one individual ligand might interact with a typical metal centre (a 'fragment orbital' approach). The departing ligand takes its lone pair of electrons with it, leaving a vacant (Lewis acidic) sp^3d^2 orbital directed towards a vacant coordination site, flanked in the d^6 case by two occupied t_{2g} orbitals of π-symmetry with respect to the M–L vector. We will return to this hypothetical situation later when considering how the molecules CO, CH_2 and C_2H_4 might bond to such a d^6 ML_5 metal centre. Similar schemes may also be constructed for other geometries and coordination numbers; however, the d^6 octahedral case is the simplest to visualize.

The 18-electron rule loses dominance at the extremes of the transition series. At the left-hand side (Groups 3 and 4) the metal centre contributes fewer electrons to the valence count, and thus more electrons are required from ligands. This may not always be easily possible simply due to the steric pressures of accommodating more ligands, *e.g.* the metallocenes MCl_2Cp_2 (M = Ti, Zr, Hf, Nb, Ta; Cp = η-C_5H_5; see Chapter 7) are stable with <18 VE (chloride is a π-donor). This also accounts for the prevalence of π-donor ligands in isolable organometallic compounds of these metals, *i.e.* ligands capable of providing extra electrons to the overall count from 'lone' pairs. Steric factors in general provide a useful method of kinetically stabilizing coordinative unsaturation (preventing bimolecular decomposition routes, which generally require vacant coordination sites), and this approach will be encountered often.

Towards the right-hand side of the transition series, electronic factors are usually associated with coordinative unsaturation. We assumed above that it will be energetically advantageous to employ all nine metal valence orbitals. For this to be true, it is necessary that these are all of comparable energy. However, on moving from Ca to Zn there is a progressive relative increase in the energy gap between the $(n-1)d$ and the ns and np orbital energies. This separation is further increased by increasing positive charge (or oxidation state) at the metal centre. For late transition metals, the np orbitals are generally less likely to participate significantly in bonding. Furthermore, as in classical coordination chemistry, the characteristic stability associated with the square planar geometry for d^8 metal centres (high energy but vacant $d_{x^2-y^2}$ orbital) extends to organometallic examples of Groups 9 and 10.

1.3 The Bonding of Unsaturated Organic Ligands to a Metal Centre

For simple σ-ligands, *e.g.* alkyls (sp^3 hybridized carbon), there are no orbitals of appropriate π-symmetry with respect to the metal–ligand

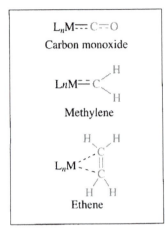

$L_nM \equiv C = O$

Carbon monoxide

$LnM = C \diagup \begin{smallmatrix} H \\ \\ H \end{smallmatrix}$

Methylene

$L_nM \diagup \begin{smallmatrix} H \quad H \\ C \\ \| \\ C \\ H \quad H \end{smallmatrix}$

Ethene

Figure 1.11 Representative π-ligands

vector and so the metal t_{2g} orbitals remain non-bonding in character. There are, however, many unsaturated organic ligands which do have orbitals of appropriate π-symmetry, be they associated with one carbon atom ['C_1': carbenes, carbynes, vinylidenes (Chapter 5), acyls (Chapter 4)] or two or more carbon atoms ['C_n': alkenes, alkynes (Chapter 6), η^n-hydrocarbons (Chapter 7)]. We will deal with three illustrative cases here: carbon monoxide (CO), methylene (CH_2) and ethene ($H_2C=CH_2$) (Figure 1.11). These basic principles can then be modified and refined in later chapters for more complex ligands.

1.3.1 Carbon Monoxide

Carbon monoxide shows very little basicity or nucleophilicity within p-block chemistry, although some Lewis acid adducts *are* known, *e.g.* $H_3B–CO$. In contrast, there are many thousands of known complexes of CO with transition metals (metal carbonyls; Chapter 3). The simple classical two-electron dative interaction from the lone pair on carbon, although important, is seldom sufficiently strong to bind CO firmly to a transition metal. Accordingly, comparatively few d^0 or d^{10} metal carbonyl complexes are known. The key lies in the presence of electrons housed in d-orbitals of π-symmetry with respect to the metal–carbon vector (t_{2g}). We begin by considering the frontier orbitals of CO (Figure 1.12).

The HOMO (3σ) is primarily based on carbon, and the degenerate set of antibonding π* orbitals (2π LUMO) also have their greatest contribution from carbon p-orbitals. The most important component of the bonding involves the combination of occupied metal t_{2g} orbitals with the

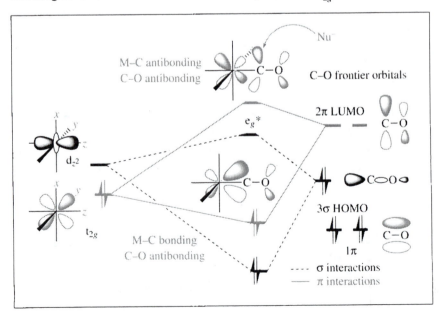

Figure 1.12 Synergic bonding of CO to a transition metal; implications for nucleophilic (Nu⁻) attack

empty π^* antibonding (2π) orbitals of CO. The interaction of the 3σ and 2π orbitals of CO with those of a transition metal is illustrated in Figure 1.12. Salient features may be summarized as follows:

(i)   The conventional two-electron dative metal–ligand σ-bond (3σ + sp^3d^2 hybrid, or d_{z^2}).

(ii)   Two orthogonal bonds of π-symmetry (e.g. d_{xz} + $2\pi_x$, d_{yz} + $2\pi_y$; only one is shown). The electrons in these bonds originated from pure metal t_{2g} orbitals but now reside in molecular orbitals with considerable ligand character. Thus a drift of electron density has occurred from the metal to the antibonding 2π orbitals of CO. The more effective is this drift of electron density from the metal to CO, the more the CO bond will be weakened, as the metal–carbon bonding increases.

This phenomenon has a number of names including π-acceptance, retrodonation or back-bonding, and is in essence similar to the C–N multiple bonding found in organic amides, in which the lone pair of an sp^2 hybridized nitrogen mixes with the empty π^* orbital of the CO bond. Ligands capable of such interactions (and there are many) are thus referred to as π-acids (in the Lewis acidity sense) or π-acceptors. The operation of these two mutually supportive interactions in opposite directions is referred to as synergic bonding. Figure 1.13 shows how we might represent this in valence bond terms.

Effective retrodonation is most readily apparent as a decrease in the stretching frequency (v_{CO}) of the CO ligand observed using vibrational (infrared, IR) spectroscopy. Free CO absorbs at 2143 cm^{-1}, but in transition metal complexes the v_{CO} frequency is typically in the range 1840–2120 cm^{-1}. Table 1.2 illustrates the effect of the π-basicity of the metal centre on v_{CO} for various metal carbonyls. Increasing the negative charge on the metal centre raises the energy of the metal t_{2g} orbitals,

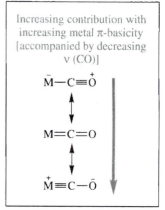

Increasing contribution with increasing metal π-basicity [accompanied by decreasing v (CO)]

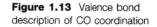

Figure 1.13 Valence bond description of CO coordination

Table 1.2 IR data (cm^{-1})^a for [M(CO)$_x$]^n

O_h-[M(CO)$_6$]^n			T_d-[M(CO)$_4$]^n		
M	n	v(CO)	M	n	v(CO)
V	1–	1859	Cr	4–	1460
V	0	1973	Mn	3–	1670
Cr	0	2000	Fe	2–	1761
Mo	0	2004	Os	2–	1787
W	0	1998	Co	1–	1883
Mn	1+	2096	Ni	0	2058
Fe	2+	2203			
Ir	3+	2254			

^{a}For CO(g), v(CO) = 2143 cm^{-1}; for [CO]$^+$(g), v(CO) = 2214 cm^{-1}

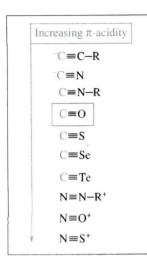

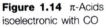

Figure 1.14 π-Acids isoelectronic with CO

making the transfer of electron density to the 2π orbitals of CO more effective and resulting in a decrease in the CO bond strength accompanied by an increase in the strength of the metal–carbon bond. Conversely, metal centres with low π-basicity (low-energy t_{2g} orbitals, positive charge, other strong and therefore competitive π-acidic co-ligands) will have labile CO ligands. The value of v_{CO} may therefore be used as an indicator of the lability of CO coordination (and susceptibility towards nucleophilic attack): the higher the v_{CO} frequency, the more labile the carbonyl ligand. The frontier orbitals of CO are analogous to those of a range of other isoelectronic ligands (*e.g.* cyanide, nitrosyls), for which the same bonding model can be applied, although the relative importance of σ-donation and π-acceptance (and hence the net acceptor ability) may vary, depending on the actual energy of the orbitals involved (Figure 1.14).

1.3.2 Metal–Carbon Multiple Bonding: Methylene

Complexes involving a formal metal–carbon double bond are referred to as either carbene or alkylidene complexes. For simplicity we shall consider the orbitals of singlet methylene, which consist of a filled sp^2 orbital ('lone pair') and an empty and highly electrophilic p_z orbital (Figure 1.15).

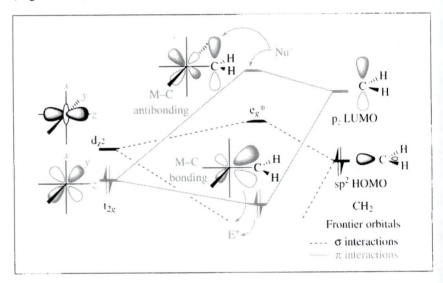

Figure 1.15 Bonding of methylene to a transition metal; implications for electrophilic (E⁺) and nucleophilic (Nu⁻) attack

Combining these with suitable valence orbitals of a transiton metal, we form a σ-bond between the sp^2 orbital on carbon and one sp^3d^2 hybrid of σ-symmetry (or d_{z^2}) on the metal centre (the classical 2VE dative σ-bond). We are now left with the vacant p_z orbital, which has correct symmetry to interact with *one* metal t_{2g} orbital (in contrast to CO, which interacts with two). This has the effect of forming a M–C π-bond, resulting in retrodonation to the carbene carbon. In general, carbenes are weaker net π-acids than CO since there is only one acceptor orbital, although this is often of lower energy than the 2π of CO. Many carbene complexes involve heteroatom carbene substituents also capable of π-donation into the empty p_z orbital, and this occurs at the expense (lengthening) of the M–C π-bond (Figure 1.16).

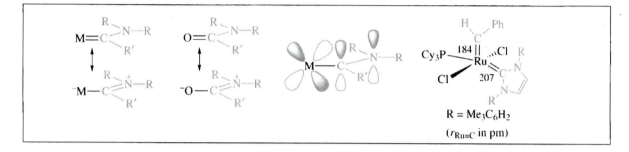

Figure 1.16 Carbene heteroatom substituents

This general description can be applied to other C_1 ligands capable of forming multiple bonds to transition metals, *e.g.* carbynes (alkylidynes), $L_nM≡CR$, and vinylidenes, $L_nM=C=CR_2$ (Chapter 5). Furthermore, a number of one-electron σ-organyl ligands, *e.g.* acyl, vinyl (alkenyl) and even aryl, have suitably vacant orbitals for such retrodative processes (Figure 1.17). It is noteworthy that the presence of a metal–carbon π-bond introduces an electronic barrier to rotation of the carbene (or other sp^2-hybridized C_1 ligand) about the metal–carbon axis (in addition to steric factors). Rotation would require that the π-component is lost at some point during rotation (Table 1.3)

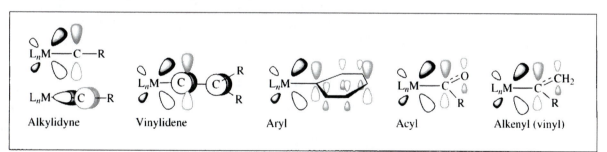

Alkylidyne   Vinylidene   Aryl   Acyl   Alkenyl (vinyl)

Figure 1.17 C_1 ligands with a π-component

This general picture of a C_1 ligand multiply bound to a metal centre allows us to predict some aspects of the reactivity of such ligands towards electrophiles and nucleophiles. The LUMO of the complex corresponds to the antibonding combination of p_z and a metal t_{2g} orbital. This is primarily localized on the carbene carbon and it is therefore this centre that is generally attacked by nucleophiles (Chapter 5). This assumes that the metal centre is coordinatively saturated (18VE); otherwise, the

Table 1.3 Barriers to carbene rotation

Complex[a]	$\Delta G^{\ddagger}$/kJ mol^{-1}	Complex	$\Delta G^{\ddagger}$/kJ mol^{-1}
Re(=CHR)(CBu^t)(OBu^t)$_2$	97.5	W{=C(NMe)$_2$C$_2$H$_4$}I$_2$(CO)$_4$	60.6
Ta(=CHPh)(CH$_2$Ph)Cp$_2$	80.3	[Fe(=CH$_2$)(dppe)Cp]$^+$	43.5
Ta(=CHR)ClCp$_2$	70.3	W{=C(NMe)$_2$C$_2$H$_4$}$_2$(CO)$_4$	43.5
W(=CHR)(NAr)(OAr)$_2$	67.8	W(=CH$_2$)Me$_3$Cp*	40.6
Nb(=CHR)ClCp$_2$	65.2		

[a] R = Bu^t; Ar = C$_6$H$_3$Pr$^i_2$; Cp = η-C$_5$H$_5$; Cp* = η-C$_5$Me$_5$; dppe = Ph$_2$PCH$_2$CH$_2$PPh$_2$

metal centre itself presents a potential site for attack by nucleophiles. Similarly, it is the carbon atom of coordinated CO that is attacked by nucleophiles (Chapter 3), a process which is favoured by metal centres of low π-basicity (low-energy t_{2g} orbitals arising from positive charge and/or oxidation state or π-acidic co-ligands). The site of attack by electrophiles is likely to be the metal centre since the HOMO has the greatest contribution from the metal t_{2g} orbital(s). In the case of carbonyl complexes, this is *almost* exclusively the case. The only departures from this involve the use of very hard [in the hard and soft acid and base (HSAB) sense] and sterically congested electrophiles, both of which will favour attack at the (hard) oxygen atom of the CO ligand (Chapter 3). It is one of the exciting aspects of organotransition metal chemistry that we can 'tune' in a controlled manner the π-basicity of the metal centre (charge, oxidation state, nature of co-ligands), and thereby control the efficacy of attack at unsaturated ligands by nucleophiles or electrophiles.

1.3.3 C$_2$–C$_n$ Ligands: Ethene

The description of the bonding of ethene to transition metals is known as the Dewar–Chatt–Duncanson model. This builds on the previous descriptions for CO and CH$_2$ with the exception that we must now consider the frontier orbitals of the ligand to be molecular orbitals delocalized over the two (or more) donor atoms. The basic features are presented for ethene; however, the bonding scheme applies in principle to the side-on coordination of any multiple bond to a transition metal.

Transition metals bind to the face of an alkene. Therefore the orbitals of interest are the filled 1π (bonding) orbital and the empty 2π (antibonding) orbital of the C=C double bond (Figure 1.18). The actual energies of these orbitals will depend on the nature of the alkene substituents, a point we will return to in Chapter 6. The π-bonding orbital (1π) has σ symmetry with respect to the metal–ligand vector and provides the electron pair for the classical dative bond. This leaves the vacant 2π (C=C antibonding) orbital available and correctly positioned for interaction with *one* filled t_{2g} orbital. Once again we see that σ-donation is accompanied by π-retrodonation in a synergic manner. In contrast to CO, the σ interaction may be sufficient to bind an alkene to a metal centre in the absence of retrodonation.

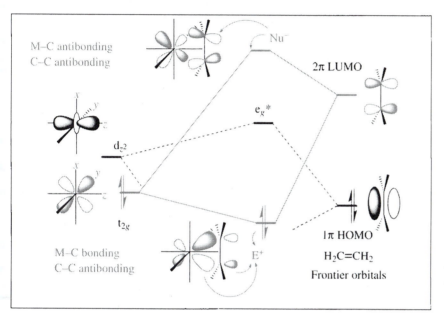

Figure 1.18 Bonding of ethene to a transition metal centre; implications for electrophilic (E⁺) and nucleophilic (Nu⁻) attack

The relative importance of the σ-donation and π-acceptance processes will be determined by the electronic nature of the metal centre and alkene substituents. This affects the reactivity of the coordinated alkene, which may be quite different from that of the free alkene. Thus, coordination of alkenes to π-basic metal centres (effective retrodonation) activates them towards electrophilic attack, while coordination to poorly π-basic metal centres activates the alkene towards nucleophilic attack. As in the case of CO coordination, π-retrodonation from the metal centre occurs into an antibonding orbital, and the effect of this is to weaken the C=C double bond. This is illustrated by structural data for the alkene complexes shown in Table 1.4. The C–C bond length in Zeise's salt, $K[Pt^{II}Cl_3(\eta\text{-}H_2C=CH_2)]$, is shorter than that in the zerovalent (more electron rich or π-basic) platinum complex $Pt^0(\eta\text{-}H_2C=CH_2)(PPh_3)_2$. Similarly, strongly electron-withdrawing alkene substituents lower the energy of the alkene 2π orbital, favouring retrodonation, as illustrated in the complexes $Fe(\eta\text{-}X_2C=CX_2)(CO)_4$ (X = H, F) for which the $F_2C=CF_2$ ligand is more strongly bound than is $H_2C=CH_2$, at the expense

Table 1.4 Selected physical data for alkene complexes[a]

Complex[b]	Rotation barrier, $\Delta G^{\ddagger}$/kJ mol^{-1}	Complex	Bond length a/nm	Bond length b/nm
Ta(H$_2$C=CH$_2$)ClCp$_2$	38.9	Nb(H$_2$C=CH$_2$)(CH$_2$R)Cp$_2$	0.230	0.141
Cr(H$_2$C=CH$_2$)(CO)(NO)Cp	47.7	Ta(H$_2$C=CH$_2$)(=CHR)(PMe$_3$)Cp*	0.226	0.148
Mn(H$_2$C=CH$_2$)(CO)$_2$Cp	35.5	W(H$_2$C=CH$_2$)(PhC≡CPh)Cl$_2$(PMe$_3$)$_2$		
Mn{C$_2$H$_2$(CO$_2$Me)$_2$}(CO)$_2$Cp	49.0	H$_2$C=CH$_2$	0.225	0.140
[Fe(H$_2$C=CH$_2$)(CO)$_2$Cp]$^+$	32.2	PhC≡CPh	0.203	0.133
[Ru(H$_2$C=CH$_2$)(CO)$_2$Cp]$^+$	31.8	Fe(H$_2$C=CH$_2$)(CO)$_4$	0.212	0.146
Fe(H$_2$C=CHCO$_2$Et)(CO)$_4$]	44.8	Fe(F$_2$C=CF$_2$)(CO)$_4$	0.199	0.153
Ru(H$_2$C=CHCO$_2$Et)(CO)$_4$]	50.6	Fe(H$_2$C=CHCN)(CO)$_4$	0.210	0.140
[Os(H$_2$C=CH$_2$)(CO)(NO)(PPh$_3$)$_2$]$^+$	44.8	[Os(H$_2$C=CH$_2$)(HC≡CH)(en)$_2$]$^{2+}$		
Rh(H$_2$C=CH$_2$)$_2$(acac)	51.3	H$_2$C=CH$_2$	0.218	0.135
Rh(H$_2$C=CH$_2$){C$_2$(OMe)$_4$}(acac)		HC≡CH	0.216	0.120
H$_2$C=CH$_2$	61.9	Rh(H$_2$C=CH$_2$)$_2$(acac)	0.213	0.121
C$_2$(OMe)$_4$	77.4	Rh(H$_2$C=CH$_2$)(F$_2$C=CF$_2$)(acac)		
Rh(H$_2$C=CH$_2$)$_2$Cp	62.8	H$_2$C=CH$_2$	0.219	0.142
Rh(H$_2$C=CH$_2$)(F$_2$C=CF$_2$)Cp		F$_2$C=CF$_2$	0.201	0.140
H$_2$C=CH$_2$	56.9	Ir(η^2-C$_{60}$)Cl(CO)(PPh$_3$)$_2$	0.219	0.153
F$_2$C=CF$_2$	–^{c}	Pt(H$_2$C=CH$_2$)(PPh$_3$)$_2$	0.211	0.143
Pt(H$_2$C=CH$_2$)$_2$(PPh$_3$)	42.7	Pt(H$_2$C=CH$_2$)$_2$(F$_2$C=CF$_2$)		
		H$_2$C=CH$_2$	0.225	0.136
		F$_2$C=CF$_2$	0.197	0.144
		Pt(H$_2$C=CH$_2$)Cl$_3$]$^-$	0.213	0.137

^{a}1 cal = 4.184 J; 1 nm = 10 Å. ^{b}acac = acetylacetonate (pentane-2,4-dionate); en = 1,2-diaminoethane (ethylenediamine). ^{c}Decomposition occurs prior to onset of rotation

of the C=C multiple bonding in the former. As in the case of carbenes (alkylidenes) above, the presence of a π-component in the bonding introduces a barrier to rotation since the π-bond must normally be broken at some point during rotation (Table 1.4). This is reflected in the rotation barriers measured for the rhodium complex Rh(η-CH$_2$=CH$_2$)-(η-CF$_2$=CF$_2$)Cp using variable-temperature NMR spectroscopy. The weakening of the C=C bond which results from retrodonation may also be inferred from IR spectroscopy, since coordination of the metal to the alkene face lowers the symmetry of the alkene, allowing the otherwise IR-inactive $\nu_{C=C}$ vibration to be observed.

We have discussed alkene coordination from a molecular orbital point of view. However, many of the conclusions also follow from a valence bond description comprising the two canonical forms shown in Figure 1.19. The nature of the metal and alkene substituents will dictate the relative contributions of the dative (1.19b) and metallacyclopropane (1.19c) canonical forms. Both molecular orbital and valence bond treatments require a pyramidalization of the trigonal carbon centres, equivalent to the hybridization at carbon assuming more sp^3 character. The bending back of alkene substituents therefore also provides an indication of the strength of the alkene coordination.

The Dewar–Chatt–Duncanson model is equally appropriate for describing the coordination of bidentate non-conjugated dienes, which may simply be viewed as two separate metal–alkene interactions, assisted kinetically by the chelate effect. The situation with conjugated dienes (trienes, *etc.*) is essentially similar with the exception that the delocalized nature of the polyene orbitals may introduce subtleties obscured by a simple valence bond description. Thus, for example, it is generally found that conjugated dienes bind more effectively than non-conjugated dienes. Indeed, there are examples of transition metal complexes that are capable of isomerizing non-conjugated dienes to their conjugated isomers (Chapter 6).

1.4   Thermodynamics of the Metal–Carbon Bond

The apparent instability of transition metal σ-organyls is now recognized to be almost invariably of kinetic origin. Indeed, the various routes by which these compounds may decompose under mild conditions actually contribute to both their intrigue and their involvement in technologically useful stoichiometric or catalytic applications. The delicate nature of some transition metal alkyls (and aryls) arises from the availability of low-energy decomposition routes that are generally less favourable for σ-organyls of the p-block elements (Chapter 4). Many of these decomposition routes may be excluded by incorporation of suitable co-ligands, which either sterically block metal coordination sites and/or confer coordinative saturation on the metal centre.

There is a general lack of *directly* comparable thermochemical data for main group and transition metal alkyls and aryls because of the very nature of the two classes of compounds. Homoleptic alkyls (*i.e.* all ligands identical) of the p-block with the exception of Group 13 show little if any Lewis acidity, and hence tend not to coordinate further ligands. The majority of isolable transition metal alkyls, on the other hand, are typically stabilized by the incorporation of kinetically or thermodynamically stabilizing co-ligands, thereby clouding direct thermochemical comparisons. The most common comparisons of thermodynamic data have involved Group 4 alkyls, which is perhaps unfortunate in that these typically have d^0 metal centres. Nevertheless, these data do confirm that transition metal–carbon σ-bonds can be quite strong.

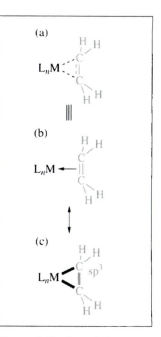

Figure 1.19 Canonical (resonance) forms for ethene coordination

There is nothing inherently unstable about single, let alone multiple, bonds between carbon and transition metals.

Bond dissociation enthalpies (BDEs; see Box 1.1) for transition metal σ-organyls appear at present to fall within the quite wide range of 70–330 kJ mol^{-1}, *i.e.* a range comparable to those of main group alkyls. It should be noted, however, that transition metals also have the possibility of including a π-component when the carbon is sp^2 (alkylidenes, alkenyls, acyls) or sp hybridized (alkylidynes, vinylidenes, alkynyls), thereby leading to larger BDEs.

In recent times it has been established that relativistic effects may enhance both σ-donation and π-acceptance for the 4d and especially the 5d transition elements. Loosely speaking, the high nuclear charge associated with heavier elements requires that the associated electrons have increasingly higher angular momentum to avoid crashing into the nucleus. This requires electronic velocities that begin to approach the speed of light, and hence the attendant increase in electronic mass leads to a decrease in the size of the corresponding Bohr radius. This is

---

Box 1.1 Bond Dissociation Enthalpies

Table 1.5 collects some representative BDEs for transition metal alkyls and aryls from which some recurrent trends emerge:

- BDEs increase down a particular group, in contrast to p-block elements

- α-Branched and cycloalkyl ligands appear less robust than *n*-alkyls

- Aryls, acyls, vinyls and especially carbenes, *i.e.* organyls based on sp^2 hybridized carbon, have higher BDEs than alkyls

- The geometry and the nature of co-ligands affects the strength of the M–C bond

- Electronegative α-substitutents (cyano, fluoro) stabilize the M–C bond

- M–H bonds are generally stronger than metal–alkyl bonds in related compounds

- The benzyl–metal bond, M–CH$_2$Ph, appears particularly weak (*cf.* the 'naturally selected' adenosyl group of vitamin B$_{12}$)

The above generalizations are of thermodynamic origin. We will see in Chapter 4, however, that kinetic factors assume prime importance in dictating the longevity of transition metal σ-organyls.

Table 1.5 M–C bond dissociation enthalpies (kJ mol^{-1})[a]

			R =	Ph	Me
CMe_4	358				
$SiMe_4$	311	Cp_2TiR_2		332	298
$GeMe_4$	249	$Cp^*_2TiR_2$		280	281
$SnMe_4$	217	Cp_2ZrR_2		300	285
$PbMe_4$	152	$Cp^*_2ZrR_2$			284
		$Cp^*_2HfR_2$			306
$Ti(CH_2Bu^t)_4$	198	Cp_2MoR_2		166	
$Zr(CH_2Bu^t)_4$	249	Cp_2WR_2		221	
$Hf(CH_2Bu^t)_4$	266	Cp^*ThR_2			339
$Zr(CH_2Ph)_4$	263	Cp^*UR_2			300
$Zr(CH_2SiMe_3)_4$	310				
Cp^*HfMe_3	294	$TaMe_5$			261
$CpPtMe_3$	163	WMe_6			160

			R =	Me	o-tolyl
Vitamin B_{12}-adenosyl	112–131				
$(RC_5H_4N)(dmg)_2CoCHMePh$	89–96	cis-$(Et_3P)_2PtR_2$		269	298
$(R_3P)(dmg)_2CoCHMePh$	73–100	cis-$(Et_3P)_2ClPtR$		251	300
$(R_3P)(dmg)_2CoCH_2Ph$	96–128	trans-$(Et_3P)_2PtPh_2$			159
$(R_3P)(oep)CoCH_2Ph$	100–124	trans-$(Et_3P)_2ClPtEt$			206
Vitamin B_{12}-R	79–104				
(R = Pr^i, CHMeEt, CH_2Bu^t, CH_2Ph)		$Cp(CO)_3MoR$			

$(CO)_5MnR$			R =	Me	203
R = H	245			Et	185
Me	187			cyclo-C_3H_5	147
Ph	207			CH_2Ph	154
CH_2Ph	129				
CH_2F	139	$Cp(CO)_3MH$			
CHF_2	144		M =	Cr	258
CF_3	182			Mo	273
C(O)Me	185			W	339
C(O)Ph	127				
$(CO)_5ReMe$	220	$(CO)_5W=C(OMe)Ph$			359
$Cp^*(PMe_3)HIrR$		$Cp^*(PMe_3)IrR_2$			
R = Ph	321		R =	H	310
cyclo-C_5H_9	215			Me	243
C_5H_{11}	244	$Cp^*(PMe_3)BrIr-CH=CH_2$			326

[a]oep = octaethylporphyrin, dmg = dimethylglyoximate. For further reading, see J. A. M. Simoes and J. L. Beachamp, *Chem. Rev.*, 1990, **90**, 629.

particularly true for electrons in s-orbitals, which have finite nuclear penetration (p- and d-orbitals have nodal planes passing through the nucleus). This results in a stabilization of the s-orbitals, which in turn more effectively shield the nuclear charge for p- and d-orbitals; these are accordingly destabilized. In our model of synergic bonding above, the ramifications are that the metal centre becomes a stronger σ-acceptor (lowering of the energy of the sp^3d^2 acceptor orbital which has some s character) and a stronger π-retrodonor [raising of the occupied π-retrodonor orbital(s)]. Thus retrodonation increases in efficiency, most significantly for the 5d metals

1.5   Accessing the Organometallic Literature

The field of organotransition metal chemistry has grown enormously in the last four decades. This reflects the natural transition from laboratory curiosities of primarily aesthetic and fundamental interest, to the routine application of organometallic reagents (stoichiometric or catalytic) in fine and bulk chemical synthesis, and their use as precursors for both trivial and sophisticated materials with wide-reaching technological applications. Just as the field has expanded and diversified, so have the media in which the science is published. The primary literature includes general chemical journals and also specialist journals dealing with organometallic chemistry [*Organometallics, J. Organomet. Chem.*]. Increasingly, organometallic chemistry is to be found in the journals of other disciplines, including those dealing with organic synthesis, polymers and the chemistry of materials. Fortunately, the secondary literature (reviews, monographs) regularly collates organometallic chemistry from these disparate sources. The most useful access points in this respect are:

Dictionary of Organometallic Compounds (Chapman and Hall). Although not comprehensive, this regularly supplemented series now available in electronic form arranges compounds firstly according to central metal and secondly according to molecular formula, providing the quickest access to the literature of a single known compound.

Comprehensive Organometallic Chemistry (Pergamon Press). The first edition covers the literature up to 1982. The second edition is supplemental and covers the literature from 1982 to 1994. The majority of material is arranged in chapters, firstly according to element and secondly according to the nature of the ligand (generally the hapticity). In addition to the element-based volumes, a number of special topics are included which span the transition series, *e.g.* transition metals in organic synthesis. The final volumes include useful cumulative indexes according to chemical formula, subject and crystallographic studies (also in searchable databases, *e.g.* www.ccdc.cam.ac.uk).

Advances in Organometallic Chemistry (Academic Press). Reviews specific aspects of organometallic chemistry and is published approximately annually [Cumulative index Volume 45].

Convenient and efficient electronic retrieval of information from chemical databases *via* the world wide web is rapidly evolving. Various services are available for retrieving information from the organometallic literature according to title, keyword, compound, author or journal. The entire literature is covered by *Chemical Abstracts* (info.cas.org); the literature from 1981 is also accessible *via* the *Web of Science* (currently available free in all British Universities: wos@mimas.ac.uk).

Finally, in addition to the many monographs on various aspects of organotransition metal chemistry, there are a number of excellent alternative textbooks, indicated in the further reading section below.

Further Reading

1. C. J. Jones, Basic Coordination Chemistry and Ligand Field Theory, in *d- and f-Block Chemistry*, RSC Tutorial Chemistry Texts 4, Royal Society of Chemistry, Cambridge, 2001.
2. Ch. Elschenbroich and A. Salzer, *Organometallics. A Concise Introduction*, 2nd edn., VCH, Weinheim, 1992.
3. M. Bochmann, *Organometallics 1. Complexes with Transition Metal-Carbon σ-Bonds*, Oxford University Press, Oxford, 1994.
4. M. Bochmann, *Organometallics 2. Complexes with Transition Metal-Carbon π-Bonds*, Oxford University Press, Oxford, 1994.
5. J. P. Collman and L. S. Hegedus, *Principles and Applications of Organotransition Metal Chemistry*, University Science Books, Sausalito, California, 1980; J. P. Collman, L. Hegedus, J. R. Norton and R. G. Finke, *Principles and Applications of Organotransition Metal Chemistry*, 2nd edn., University Science Books, Sausalito, California, 1987.
6. F. Mathey and A. Sevin, *Molecular Chemistry of the Transition Elements*, Wiley, Chichester, 1996.

Sections on organotransition metal chemistry are included in general textbooks on inorganic chemistry, including:

7. F. A. Cotton, G. Wilkinson, C. A. Murillo and M. Bochmann, *Advanced Inorganic Chemistry*, 6th edn., Wiley-Interscience, New York, 1999.
8. D. F. Shriver and P. W. Atkins, *Inorganic Chemistry*, Oxford University Press, Oxford, 3rd edn., 1998.
9. N. N. Greenwood and A. Earnshaw, *Chemistry of the Elements*, 2nd edn., Butterworth-Heinemann, Oxford, 1997.

2

Co-ligands in Organotransition Metal Chemistry

Aims

- By the end of this chapter you should be familiar with some of the more frequently encountered co-ligands in organometallic chemistry. These include phosphines, nitrosyls, (spectator) cyclopentadienyls, carbaboranes, calixarenes, silsesquioxanes, corrins and porphyrins
- Not all of these are 'innocent' spectators; some take part in, or influence indirectly, the course of reactions of the metal–carbon bond
- Among these, the hydride ligand is ubiquitous in organometallic chemistry. Accordingly, this simplest conceivable ligand is considered in detail

2.1 'Innocent' Co-ligands

Throughout the text, when the co-ligands are of only secondary interest, or when the generality of a reaction is to be emphasized, the abbreviation L_nM will be used to indicate a general metal 'M' with a set of 'n' ligands. Ligands which adopt a purely spectator role are often referred to as 'innocent'.

2.1.1 Phosphines: Electronic, Steric and Chirality Control

Aliphatic *(i.e.* sp^3 N) amines have a rich coordination chemistry dating back to the seminal studies of Alfred Werner, which raised the curtain on coordination chemistry.

Alfred Werner was awarded the Nobel Prize for chemistry in 1913.

They are strong σ-donors, but have no energetically accessible molecular orbitals available of the correct π-symmetry for retrodative combination with occupied t_{2g} type metal orbitals (although aromatic imines, *e.g.* 2,2'-bipyridyl, may do so). Phosphines, PR_3, their heavier analogues (and, to a lesser extent, arsines and stibines) present a quite different situation. These molecules have empty orbitals of π-symmetry with respect to the metal–ligand bond. In principle, vacant pure phosphorus d-orbitals have suitable symmetry for constructive overlap with occupied metal t_{2g} orbitals (Figure 2.1a). Alternatively, the σ-bonds, which bind the substituents to phosphorus, have vacant antibonding (σ^*) partners which also have local π-symmetry with respect to the M–P vector (Figures 1.2c and 2.1b). As the electronegativity of the phosphorus substituents increases (alkyl < aryl < alkoxide < halide), the energy of the phosphorus d-orbitals and the σ^* orbitals will decrease, thereby increasing the facility of any retrodative (π-acceptor) interaction from the metal.

Thus the first important feature of phosphines emerges: the σ-dative/π-retrodative capacity of phosphine ligands may be fine-tuned to allow control over the electronic nature of the metal centre. A quantitative indication may be derived from the complexes $Ni(CO)_3(PR_3)$. This range of (C_{3v} symmetric) complexes gives rise to a single intense IR absorption (a_1 ν_{CO}) in the region 2050–2110 cm^{-1}, dependent upon the substituents 'R' (Table 2.1). At one extreme is the highly basic and bulky phosphine PBu^t_3, for which retrodonation is negligible. The net effect of the various substituents 'R' is approximately additive, leading to the empirical expression given in Table 2.2 based on the substituent contributions X_i.

The second variable feature of phosphines is the steric profile. An approximate measure of this is given by the Tollman cone angle (θ, Figure 2.2). This angle is derived by considering the cone which would be defined by the van der Waals surface generated by a phosphine bound

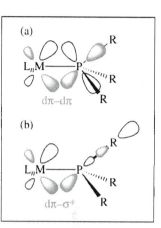

Figure 2.1 Retrodonation to phosphines

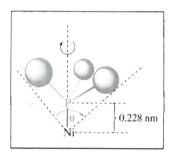

Figure 2.2 Steric properties of phosphines (cone angle)

Table 2.1 Steric (θ) and electronic (ν) properties for phosphines

PR_3	$\theta/°$	ν/cm^{-1}	PR_3	$\theta/°$	ν/cm^{-1}
PBu^t_3	182	2056	$P(OMe)Ph_2$	132	2072
PCy_3	170	2056	$PHPh_2$	128	2073
$P(C_6H_4OMe)_3$	194	2058	$P(OPr^i)_3$	130	2076
PPr^i_3	160	2059	$P(OEt)_3$	109	2076
PBu_3	132	2060	PH_2Ph	101	2077
PEt_3	132	2062	$P(OMe)_3$	107	2080
$P(NMe_2)_3$	157	2062	PH_3	87	2083
PMe_3	118	2064	$P(OPh)_3$	128	2085
PMe_2Ph	122	2065	$P(C_6F_5)_3$	184	2090
$P(CH_2Ph)_3$	165	2066	PCl_3	124	2097
$PMePh_2$	136	2067	PF_3	104	2111
PPh_3	145	2069			

Table 2.2 Phosphine substituent contributions for Ni(CO)$_3$(PR$_3$): $v_{CO} = 2056.1 + \Sigma_i X_i$

R	X_i	R	X_i	R	X_i
Bu^t	0.0	Me	2.6	H	8.3
Cy	0.1	Ph	4.3	OPh	9.7
Pr^i	1.0	OPr^i	6.3	C$_6$F$_5$	11.2
Bu	1.4	OEt	6.8	Cl	14.8
Et	1.8	OMe	7.7	F	18.2
NMe$_2$	1.9				

to nickel which freely rotates about the Ni–P bond (assuming $r_{Ni–P} = 0.228$ nm). Table 2.1 collates the Tollman cone angle for a range of phosphines, revealing the impressive control over steric and electronic properties offered. The choice of nickel(0) is not entirely arbitrary. The hydrocyanation of butadiene is catalysed by nickel phosphine complexes, providing adiponitrile [N≡C(CH$_2$)$_4$C≡N], which is a key intermediate in the preparation of 6,6-Nylon (Chapter 6). An understanding of the electronic and steric properties of the nickel catalysts guided the process and catalyst optimization.

Unlike amines, which readily invert their configuration at nitrogen, phosphines have substantially larger barriers to inversion (NH$_3$, 25 kJ mol^{-1}; PH$_3$, 155 kJ mol^{-1}). Resolved chiral phosphines therefore retain their configuration at phosphorus and this chirality may influence the enantioselectivity of reactions involving their complexes. Chiral organometallics are often capable of mediating enantioselective transformations of substrates, most usefully in a catalytic manner (Chapter 6). In recent times, effort has focused more on the use of phosphines bearing chiral substituents. Some examples of technologically useful chiral phosphines are shown in Figure 2.3 and are revisited in Chapter 6.

Finally, just as polydentate amines find wide application in biological and synthetic coordination chemistry, so too do bi- and to a lesser extent tridentate phosphines. In addition to simple kinetic stabilization through

Figure 2.3 Selected chiral phosphines

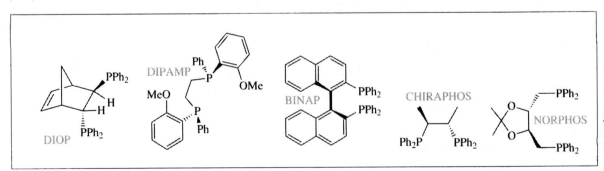

the chelate effect, the ring size of chelates formed between transition metals and bidentate phosphines is becoming recognized as a key factor in catalyst design. The bite angle of bidentate phosphines (α, Figure 2.4) can have an impact on the hybridization at the metal centre and the resulting energies of valence orbitals. Furthermore, steric factors associated with increased chelate size may contribute pressures on ligands undergoing conversions, *e.g.* reductive elimination and migratory insertion (see Chapter 3). Bis(phosphino)methanes, $H_2C(PR_2)_2$, present a special case where coordination across a metal–metal bond can favour the formation of binuclear complexes, sometimes in preference to mononuclear structures which require the formation of more strained four-membered metallacycles (Figure 2.5).

Phosphines find wide application as co-ligands in industrial catalysis. In addition to their use in conventional organic solvents, variants exist to make such complexes amenable to more exotic (and environmentally friendly) chemical environments, *e.g.* aqueous–organic two-phase systems (Figure 2.6a), fluorophasic solvents (Figure. 2.6b) or immobilized on polymer supports (Figure 2.6c). Each of these variants attempts to overcome the problem of ultimate catalyst separation (and re-use) from products, the advantages being both economic and environmental.

Arsines (AsR_3) and stibines (SbR_3) are generally found to be more labile (sometimes an advantage) than phosphines, in part due to their reduced σ-donor properties. Furthermore, arsenic and antimony lack conveniently measurable spin-active nuclei for NMR studies, in contrast to phosphines (^{31}P: $I = \frac{1}{2}$, 100% abundance), for which an enormous amount of informative data is available. Because of the direct bond between phosphorus and the metal centre, information is available about the electronic nature and stereochemistry of the metal centre, in addition to information about dynamic processes. This is further enhanced in the case of complexes involving metals with NMR-active nuclei, *e.g.* W, Rh, V, Pt, Hg and Ag, where $^1J(M-^{31}P)$ couplings may be observed.

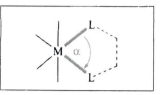

Figure 2.4 Chelate bite angle (α)

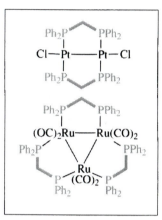

Figure 2.5
Bis(phosphino)methanes

Figure 2.6 (a) Hydrophilic, (b) fluorophasic and (c) polymer-supported phosphines

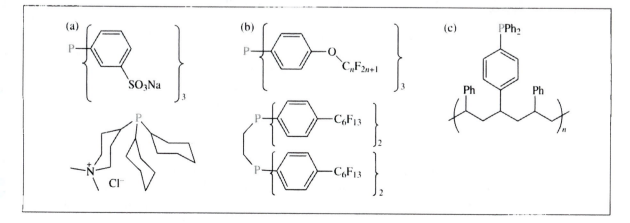

2.1.2 Aryloxides, Siloxides and Calixarenes

In contrast to phosphines which are 'soft' by nature, alkoxides and aryloxides are particularly 'hard' in the hard and soft acid and base (HSAB) sense. In addition, they carry two 'lone pairs' of electrons which may (if required) become involved in π-donation to an electron-deficient metal centre, thereby relieving coordinative unsaturation. If both pairs of electrons can be 'accepted' by the metal, then the alkoxide could be considered to serve as a net 5VE donor. Accordingly, some parallels have emerged between, for example, the organometallic chemistry of the 'TiCp$_2$' (Chapter 7) and 'Ti(OAr)$_2$' (Ar = aryl) fragments (Figure 2.7). This analogy is most useful when the alkoxide (OR) or aryloxide (OAr) has bulky substituents which disfavour dimerization, e.g. OCH(CF$_3$)$_2$ and OC$_6$H$_3$Ph$_2$-2,6 via alkoxide bridge formation. Similar arguments hold for amides (L$_n$M–NR$_2$), which can in principle act as 3VE donors when suitably vacant metal orbitals are present. Once again, large substituents may be employed to prevent dimerization, e.g. in the three-coordinate complexes M{N(SiMe$_3$)$_2$}$_3$ (M = Y, Ti, Cr, Mn, Fe, Co). Imides (L$_n$M=NR) and oxides (L$_n$M=O) can provide between 2VE and 4VE to the valence count and hence parallels between the chemistries of the 'TiCp$_2$', 'VCp(=NR)' 'VCp(=O)'and 'Cr(=NR)$_2$' fragments can be considered (Figure 2.7).

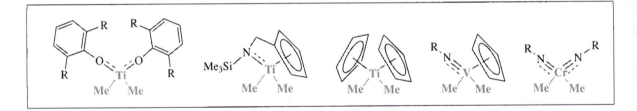

Figure 2.7 Parallels in the chemistry of cyclopentadienyl, aryloxide, amido and imido ligands

The study of organometallic alkoxides has a further impetus: many industrial heterogeneous transition metal catalysts result from the adsorption of metal salts onto the surface of various of inorganic oxides (alumina, silica, titania, etc.). In general, heterogeneous catalysts offer the practical advantages of product separation, long catalyst lifetime and re-use. The downside is that the truly active species in such systems (and there may be more than one, leading to a loss of selectivity) are very difficult to observe directly using conventional spectroscopic techniques. Such a catalyst is exemplified by the Phillips alkene polymerization catalyst obtained from the absorption and reduction of CrO$_3$ onto silica gel. The coordination number and geometry of the active species are by no means certain and even the oxidation state of the active chromium centre attracts debate (Cr^{III} or Cr^{II}?). It is possible, however, to gain insight into possible pathways by emulating the active site with soluble model complexes (*cf.* Figure 1.2), which are more amenable to spectro-scopic and kinetic study. Two ligands which offer promise in modelling the behaviour of a metal centre bound to an oxide surface are provided by calixarenes (Figure 2.8a) and silsesquioxanes (Figure 2.8b). Both of these have an arrangement of four or three oxygen donor sites, respectively, ideally disposed to coordinate metal centres and provide soluble organometallics which may be studied in more detail than the heterogeneous systems.

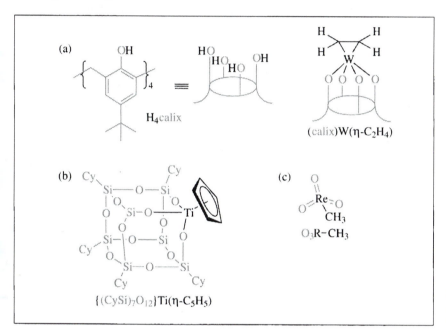

Figure 2.8 Soluble model compounds for heterogeneous oxide-bound organometallics

2.1.3 Porphyrins, Corrins and Related Macrocycles

In nature, vitamin B_{12} has evolved as a highly effective organometal-lic metalloenzyme. It is therefore not surprising that synthetic chemists have attempted to understand and harness the properties of such macro-cyclic organometallic compounds. The vitamin B_{12} nucleus involves a corrin ring, in which the circumannular electronic unsaturation is inter-rupted (Figure 1.3b). Complex derivatives of the more symmetric dian-ionic and aromatic $(4n + 2 = 20)$ porphyrin macrocycle are found in many non-organometallic metalloproteins (haemoglobin) and enzymes (cytochrome P-450). Because of the comparative ease of preparation and symmetry, simpler models of these complex biological macrocycles, *e.g.* tetraphenylporphyrin (H_2TPP, Figure 2.9a), have been studied in much more detail than the corrins. An even simpler macrocycle (H_2tmtaa, Figure 2.9b) is obtained by the (nickel-templated) condensation of pen-tane-2,4-dione (Hacac) and *o*-phenylenediamine. The tmtaa ligand is in some ways reminiscent of a porphyrin with three distinctions. The dianion ligand is anti-aromatic (16π), the pocket for metal coordination is smaller and the ring more flexible. These factors tend to favour a *cis* geometry for octahedral complexes ML$_2$(tmtaa), although some smaller metal ions, *e.g.* low-spin d^6 and d^8 centres, can fit into the plane of the macrocycle. Finally, phthalocyanine (Pc) complexes (Figure 2.9c) are known for most metals, although it is their chemical inertness rather than reactivity that underpins the majority of their technological applications, and accounts for their somewhat limited use in organometallic chemistry.

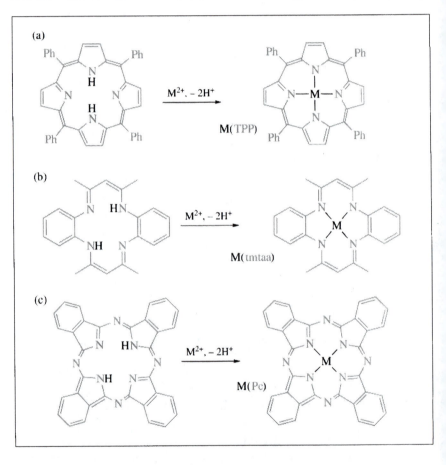

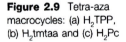

Figure 2.9 Tetra-aza macrocycles: (a) H$_2$TPP, (b) H$_2$tmtaa and (c) H$_2$Pc

2.1.4 η^n-C$_n$H$_n$ Carbocyclic Polyenes and Related Ligands

The ligand-based chemistry and metal–ligand bonding of this class of ligand will be detailed in Chapter 7. However, by far the majority of compounds involving such ligands employ them as spectators rather than direct participants. A cyclic unsaturated ligand bound in an *n-hapto* manner, η^n-C$_n$R$_n$, provides nVE to the electron count (Table 1.1). The coordination involves adjacent sites and therefore the ligands bind in a pseudo-facial manner, ensuring that the remaining ligands have mutually *cis* arrangements.

The most widely employed ligands in this class are η^6-arenes and η^5-cyclopentadienyls, providing 6VE and 5VE respectively to the EAN count. Figure 2.10 shows some of the numerous ways in which the arene or cyclopentadienyl ligands might be functionalized to introduce control over the electronic, steric and chiral properties of the metal complex. Many of these functionalizations may be carried out even after the ligand has been coordinated (Chapter 7). Note that introducing two different substituents into the cyclopentadienyl (any relative positions) or arene ligand (1,2- or 1,3-positions) renders the complex chiral (no elements of symmetry) because of the metal centre coordinating to one or other of the prochiral faces of the planar ring.

When two of these ligands are bound to one metal the complexes are referred to as **'sandwich'** complexes. If only one of these ligands is bound to a metal, the complexes are referred to as **'half-sandwich'** or **'piano stool'** complexes.

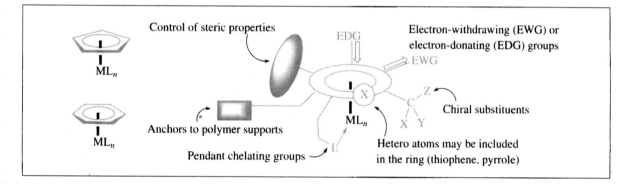

Figure 2.10 Functionalization of η^n-C_nH_n facial ligands

2.1.5 Poly(pyrazolyl)borates

The reactions of tetrahydridoborate $[BH_4]^-$ salts with variously substituted pyrazoles provide access to a wide range of poly(pyrazolyl)borate chelates (Figure 2.11). The most widely studied are the tris(pyrazolyl)borates, which typically coordinate to three *facial* sites of a complex, providing a total of 5VE to the electron count (if the ligand is considered neutral). An analogy with the coordination properties of the cyclopentadienyl ligand therefore emerges, and many analogous pairs of complexes have been prepared. One of the most useful features of this class of ligand arises from the orientation of substituetns in the 5-position which are directed towards the metal coordination sphere. This affords control over the steric profile of the metal centre. Chiral examples have also been prepared. The bis(pyrazolyl)borates have been less studied; however, these offer the added feature of possible agostic coordination (see Figure 2.25) of a B–H bond to a vacant metal coordination site.

For late transition metals, it is common for one of the pyrazolyl arms of a tris(pyrazolyl)borate to dissociate, adopting a bidentate coordination. This is especially common for the electronically favorable d^8 square planar geometry, where facial coordination is less advantageous.

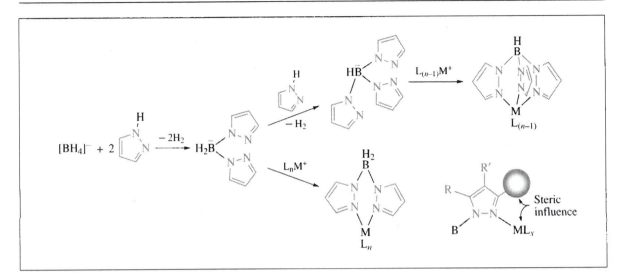

Figure 2.11 Synthesis and coordination of poly(pyrazolyl)borates

2.1.6 Carbaboranes

The fields of main group cluster chemistry and organotransition metal chemistry converge with complexes of the carbaborane ligands $[RCB_{10}H_{10}]^{3-}$ (carbollide) and $[R_2C_2B_9H_9]^{2-}$ (dicarbollide). Although these complexes themselves feature direct metal–carbon bonds, it is their analogy with cyclopentadienyl ligands that provides the key focus for their study in an organometallic context. The frontier orbitals of each of these anions extend from the open five-membered face and have the same symmetry and occupancy as those of the Cp anion (Figure 2.12). Coordination of a metal centre to this face completes the beautiful icosohedral motif. The introduction of a variety of functional groups at carbon is fairly straightforward, further increasing the possible coordination properties.

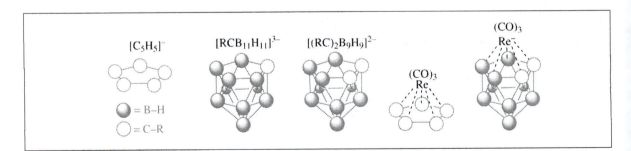

Figure 2.12 Carbaboranes: cyclopentadienyl mimics

2.2　Hydride Complexes

The field of metal hydride chemistry has developed contemporaneously with that of organometallic chemistry. The first example, cis-FeH$_2$(CO)$_4$, arose in the laboratory of Walter Hieber (Munich, 1931), from the study of reactions of metal carbonyls (Chapter 3) with bases, including hydroxide (Figure 2.13).

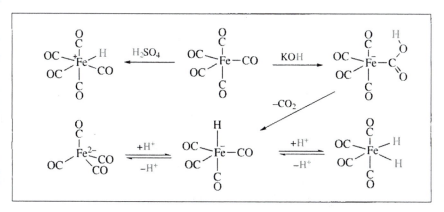

Figure 2.13 Synthesis of iron carbonyl hydride complexes

2.2.1 Characterization of Metal Hydrides

Many industrial organometallic catalytic processes involve organometallic hydride complexes as intermediates (Chapter 6). An understanding of the chemistry of these ligands is of fundamental as well as economic importance to organometallic chemists. For some time following their initial discovery there was debate as to their actual nature: were they true conventional ligands or, as initially suggested, buried protons within the valence shell of the metal centre? This debate was finally put to rest with the application of X-ray and neutron diffraction crystallographic techniques to the characterization of the complexes PtHBr(PEt$_3$)$_2$ and MnH(CO)$_5$, respectively. The scattering of X-rays by crystals is primarily by electrons. Hydrogen (one associated electron) is therefore the most difficult atom to locate accurately, especially if it is bound directly to a heavy metal. In the case of PtHBr(PEt$_3$)$_2$ the hydride ligand was not actually 'located'; instead, a T-shaped geometry was observed suggesting that a coordination site was occupied by the inconspicuous hydride ligand. Factors other than electron density contribute to neutron scattering, thereby allowing positions of hydrogen atoms to be identified, *e.g.* in MnH(CO)$_5$ (Figure 2.14), by this less readily and expensive technique.

　The salient structural features of this molecule include (i) approximate C_{4v} symmetry; (ii) a strong *trans* influence exerted by the hydride ligand

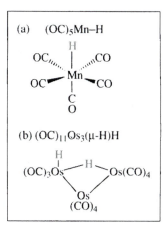

Figure 2.14 Molecular structures of metal hydrides

[the ability of a ligand to weaken (lengthen) the bond of the ligand to which it is *trans* coordinated]; (iii) bending of the *cis* ligands towards the smaller hydride ligand. Similar features are observed for the five-coordinate trigonal bipyramidal complex $CoH(CO)_4$, wherein the hydride adopts an axial coordination site. The conjugate bases of these molecules ($[Mn(CO)_5]^-$ and $[Co(CO)_4]^-$) have trigonal bipyramidal and tetrahedral geometries, respectively, *i.e.* protonation results in a change in geometry and increase in coordination number.

Terminal metal hydrides typically give rise to IR absorptions in the region 1800–2100 cm^{-1}. However, the polarity of the metal–hydrogen bond is highly variable, as are the intensities of resulting IR absorptions, *e.g.* $IrH_5(PMe_3)_2$ (intense) and $CoH\{P(OPh)_3\}_4$ (very weak), although the assignments may be verified by deuterium replacement. Bridging hydrides absorb to lower frequency (1000–1550 cm^{-1}).

^{1}H NMR spectroscopy is by far the most reliable and convenient spectroscopic technique. The majority of metal hydrides resonate to high field of $SiMe_4$, between δ –3 and –25 ppm. Hydrides of the more electropositive metals (Groups 4–6), however, have resonances to lower field (Table 2.3). Bridging hydrides typically resonate to even higher field, whilst interstitial hydrides (see later) show somewhat unpredictable behaviour, *e.g.* $[Ru_6(\mu_6\text{-}H)(CO)_{18}]^-$ (δ +16.4). In general, however, the chemical shift dependence of metal hydride resonances is complex and not easily predictable. The utility of NMR spectroscopy is further enhanced when either the metal itself has a spin-active nucleus or when co-ligands, especially phosphines, for example, have 'NMR friendly' nuclei. Two examples are illustrated in Figure 2.15, where it is shown that stereochemical information follows from the magnitude of couplings between the proton and the phosphorus nuclei.

The hydrogen atom acts as a stereochemically active ligand; the metal–hydrogen distance is that of a normal covalent bond.

For metal hydrides,
$v_{MH} = \sqrt{2} \times v_{MD}$,
e.g. ReHCp$_2$ and ReDCp$_2$ have
$v_{ReH} = 2000$ and $v_{ReD} = 1460$
cm^{-1}, respectively.

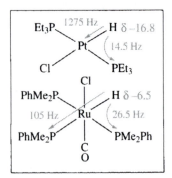

Figure 2.15 $^1J(PH)$ and $^2J(PH)$ couplings

Table 2.3 δ(^{1}H) for metal hydrides

Hydride[a]	δ(^{1}H)/ppm	Hydride[a]	δ(^{1}H)/ppm
H$_2$ZrCp*$_2$	7.46	HReCp$_2$	–12.8
H$_3$NbCp$_2$	12.6, 13.5	[HFeCp$_2$]$^+$	–2.1
H$_3$TaCp$_2$	–1.65, –3.02	[HRuCp$_2$]$^+$	–7.2
H$_2$MoCp$_2$	–8.76	HMn(CO)$_5$	–7.5
H$_2$WCp$_2$	–12.28	HRe(CO)$_5$	–5.66
[H$_3$WCp$_2$]$^+$	–6.08, –6.44	H$_2$Fe(CO)$_4$	–11.1
[H$_2$TcCp$_2$]$^+$	–14.6	H$_2$Ru(CO)$_4$	–7.62
HTcCp$_2$	–7.8	H$_2$Os(CO)$_4$	–8.84

[a]Cp = η-C$_5$H$_5$; Cp* = η-C$_5$Me$_5$

Paramagnetic metal hydrides are comparatively rare; however, when they arise, EPR (ESR) spectroscopy may be useful, e.g. the anionic complex $[TiH_2Cp_2]^-$ gives rise to a resonance (d^1, $g = 1.992$) showing coupling to the two hydride ligand nuclei ($a = 7-8$ G). The majority of metal hydrides (and all examples met in this text) are, however, diamagnetic ('EPR silent'). Mass spectrometry is very useful for establishing the gross molecular composition, so long as it is used with caution owing to the often facile loss of hydrogen atoms or molecules from ions.

In calling such complexes 'metal hydrides', an historical analogy with main group metal hydrides is implied. In terms of the chemical nature of the transition metal–'hydride' bond, however, a spectrum of behaviour is observed from 'hydridic' ($^{\delta+}M-H^{\delta-}$) through to strongly 'protic' ($^{\delta-}M-H^{\delta+}$). Figure 2.16 illustrates two reactions for the complex $ZrH_2Cp^*_2$, which are reminiscent of the behaviour of Group 13 metal hydrides (PE: Zr = 1.3; Al = 1.6), suggesting a hydridic bond polarization. In contrast, Table 2.4 compiles pK_{a1} values for selected later transition metal (PE 1.9–2.3) hydrides, which clearly illustrate Brønsted acidic behaviour (see also Figure 2.13). This series parallels to an extent the nucleophilicity of the conjugate bases obtained on deprotonation (Chapter 3).

Table 2.4 pK_{a1} for metal hydrides in acetonitrile (water)

Complex	pK_{a1}
$HCr(CO)_3Cp$	13.3
$HMo(CO)_3Cp$	13.9
$HW(CO)_3Cp$	16.1
$HMn(CO)_5$	15.1 (7.1)
$HRe(CO)_5$	21.0
$H_2Fe(CO)_4$	11.4 (4.0)
$H_2Ru(CO)_4$	18.7
$H_2Os(CO)_4$	20.8
$HCo(CO)_4$	8.4 (strong)
$HCo(CO)_3(PPh_3)$	15.4 (6.96)

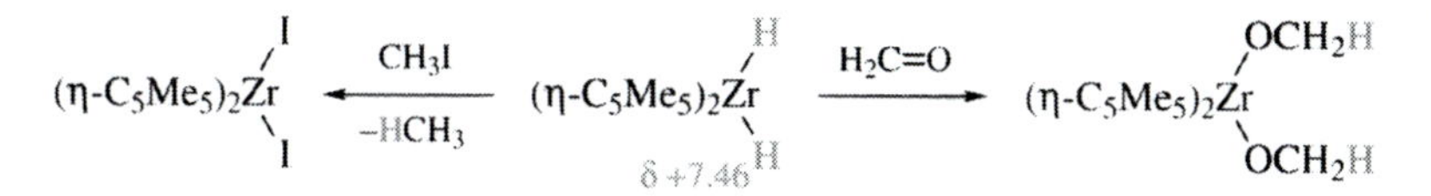

$$(\eta\text{-}C_5Me_5)_2Zr \overset{CH_3I}{\underset{-HCH_3}{\longleftarrow}} (\eta\text{-}C_5Me_5)_2Zr \overset{H_2C=O}{\longrightarrow} (\eta\text{-}C_5Me_5)_2Zr$$

Figure 2.16 Borane-like reactivity of metal hydrides

2.2.2 Bridging Hydrides

In this text we will be primarily concerned with the reactivity of terminal hydride ligands. There are, however, other types of metal hydride which deserve mention, not in the least because of their possible significance in heterogeneous industrial processes. One of the most important classes of metal hydrides is the polyhydrides, i.e. complexes with two or more hydride ligands. Until the discovery of dihydrogen complexes (see below) these were considered to contain only discrete terminal metal hydride ligands. Many have now been reformulated as involving coordinated dihydrogen, e.g. $RuH_4(PPh_3)_3$ is in fact $RuH_2(\eta\text{-}H_2)(PPh_3)_3$ (Figure 2.17).

Just as bridging hydrides play a central role in the chemistry of Group 13 hydrides, three centre–two electron (3c–2e) bonding is a recurrent feature of binuclear (and polynuclear) hydrides of the transition metals

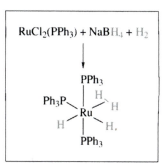

Figure 2.17 Polyhydride vs. dihydrogen coordination

(Figure 2.18), with the added dimension that the metal–metal bonds may be formally multiple, *e.g.* $Os_3(\mu\text{-H})_2(CO)_{10}$ (Os=Os), $[Ir_2(\mu\text{-H})_3Cp^*_2]^+$ (Ir=Ir) and $Re_2(\mu\text{-H})_4H_4(PEt_2Ph)_4$ (Re≡Re) or involve different metals.

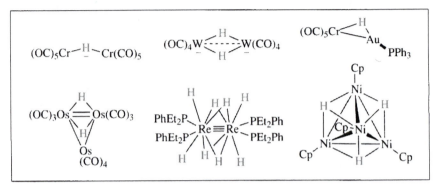

Figure 2.18 Bridging hydride complexes

The energy involved in movement between terminal and bridging positions may be modest, illustrated by the complexes $Os_3(\mu\text{-H})H(CO)_{11}$ (Figure 2.14b) and $Re_2(\mu\text{-H})_4H_4(PEt_2Ph)_4$ (Figure 2.18); each of these molecules have both terminal and bridging hydride ligands. This is further illustrated by the complex $W_2(\mu\text{-H})_2H_6Cp^*_2$, the high-field ^1H NMR spectrum of which comprises a single resonance (δ –0.34) with 'satellites' due to coupling to two ^{183}W nuclei $[J(^{183}W\text{–}^1H) = 48\ Hz]$, indicating that all eight protons see an averaged chemical environment (on the ^1H NMR timescale). Hydrides may also triply bridge the face of a tri- or polymetallic cluster.

2.2.3 The Cluster–Surface Analogy

It has been argued that the behaviour of ligands on the edges and faces of metal clusters (molecular compounds based on three or more connected metals) might model the behaviour of such ligands respectively on a step or surface of a bulk metal (the cluster–surface analogy, Figure 2.19).

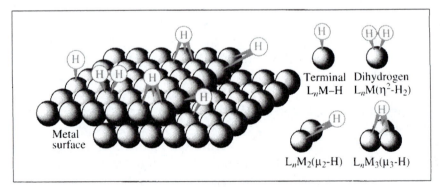

Figure 2.19 The cluster–surface analogy

Such entities are difficult to directly observe spectroscopically on a metal surface; however, molecular clusters (and their chemical transformations) are *usually* amenable to spectroscopic studies, which can shed light on the more obscure realm of surface chemistry. In this way, doubly (μ_2) and triply (μ_3) bridging hydrides can serve as conceptual models for hydrogen adsorbed on the surface of metals. Similar arguments apply to other ligands, *e.g.* carbonyls (Chapter 3), alkyls and acyls (Chapter 4), alkenes (Chapter 6) and carbenes (Chapter 5). Hydride ligands may also reside within the interstices of molecular clusters, thereby providing a conceptual link with the nature of hydrogen adsorbed within a bulk metal.

2.2.4 Dihydrogen Complexes

There are many examples of complexes with more than one hydride ligand. Until the mid-1980s these were all assumed to contain discrete hydride ligands. It was suspected that in situations where dihydrido complexes arose from the oxidative addition of dihydrogen, an intermediate coordination of the H–H bond must occur *en route* to cleavage, accounting for the commonly observed *cis* stereochemistry. Thus, for example, Vaska's complex IrCl(CO)(PPh$_3$)$_2$ reacts with I–CH$_3$ by a two-step oxidative addition process, resulting in *trans* addition of the methyl and iodide components (Figure 2.20). In contrast, the reaction with hydrogen leads to a *cis*-dihydride, IrH$_2$Cl(CO)(PPh$_3$)$_2$.

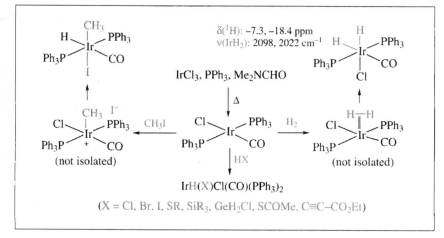

Figure 2.20 Synthesis and oxidative addition reactions of Vaska's complex

In 1984, however, Kubas succeeded in isolating the first example of a dihydrogen complex from the (reversible) reaction of W(CO)$_3$(PCy$_3$)$_2$ (16VE!) with hydrogen (Figure 2.21). This complex retains a bond between the two hydrogen atoms [$r_{H–H}$ = 0.084 nm (0.074 nm in H$_2$); $r_{W–H}$ = 0.175 nm], *i.e.* oxidative addition has been arrested. In the interim,

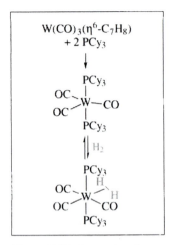

$W(CO)_3(\eta^6\text{-}C_7H_8)$
$+ 2\ PCy_3$

$\downarrow$

PCy$_3$
OC$\quad$|
$\quad$W—CO
OC$\quad$|
$\quad$PCy$_3$

$\parallel$ H$_2$

PCy$_3$
OC$\quad$H
$\quad$W$\quad$H
OC$\quad$CO
$\quad$PCy$_3$

Figure 2.21 A complex of dihydrogen

numerous examples of such complexes have since been identified, including both new complexes and reformulations of older 'polyhydride' complexes, some of which include both hydride and dihydrogen ligands (Figure 2.18).

Various spectroscopic techniques have been developed and refined to identify such situations, amongst which ^{1}H NMR spectroscopy remains the most informative. By preparing complexes of HD rather than H$_2$, H–D coupling constants $[^1J(^1\text{H}-^2\text{H})]$ may be measured. In situations where there is direct bond between the H and D atoms, then the coupling will be larger (typically *ca.* 20–30 Hz) than for a dihydride (0–3 Hz), *e.g.* 33 Hz for $W(\eta^2\text{-HD})(CO)_3(PCy_3)_2$ and 43 Hz for free HD. More conveniently, use can be made of the phenomenon that one proton moderates the rate at which a nearby proton relaxes in the NMR experiment (T_1, spin relaxation time). This effect diminishes very quickly (r^{-6}) with internuclear separation 'r'. Spin relaxation times (a routinely measured parameter with modern spectrometers) of $T_1 < 150$ ms are generally taken as suggestive of dihydrogen coordination.

For the dihydrogen molecule the frontier orbitals are simply the occupied bonding 1σ orbital and an empty antibonding 2σ* orbital. If these approach a metal side-on (perpendicular to the H–H bond axis, Figure 2.22), a synergic situation arises with donation (combination of 1σ → sp^{3}d^2) and retrodonation (t$_{2g}$ → 2σ*). Both of these processes should be mutually supportive (synergic). However, as the efficiency of each of these increases, the depletion of the 1σ and population of the 2σ* orbitals will lead ultimately to rupture of the H–H bond, *i.e.* oxidative addition. Therefore, the situations where dihydrogen complexes arise and persist involve a balance between these processes ensuing sufficiently to bind the dihydrogen molecule as a ligand, but not to the extent that oxidative addition occurs. There are various ways of discouraging the ultimate

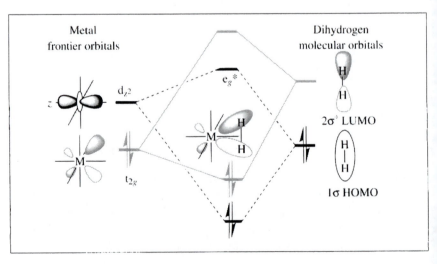

Metal frontier orbitals

Dihydrogen molecular orbitals

d_{z^2}

e_g*

z

M

M

t_{2g}

H
H
2σ^3 LUMO

H
|
H
1σ HOMO

Figure 2.22 Synergic bonding in dihydrogen coordination: hydrogen to metal donation and metal to hydrogen retrodonation

oxidative addition (lowering the energy of the π-retrodative metal orbital), *e.g.* loading the metal coordination sphere with strong π-acid co-ligands or making the complex cationic. One or both of these features typify the majority of isolated dihydrogen complexes, with the proviso that, in the case of cationic complexes, these are protected from potential bases; dihydrogen ligands can act as strong Brønsted acids to generate monohydride complexes (Figure 2.23).

The basic molecular orbital scheme shown in Figure 2.22 for the coordination of a H–H σ-bond to a transition metal is in principle applicable to the coordination of any H–element σ-bond. In particular, there are many examples of 3c–2e bonding involving coordination of C–H σ-bonds (Figure 2.24) and other heteroatoms, *e.g.* B–H, Si–H, Ge–H, N–H bonds (Figure 2.25). The term 'agostic' has been used to denote such interactions, in particular for situations involving C–H–M interactions, and these will be met subsequently.

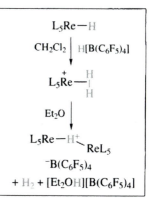

Figure 2.23 Brønsted acidity of dihydrogen ligands; $L_5Re = Re(CO)_4(PCy_3)$

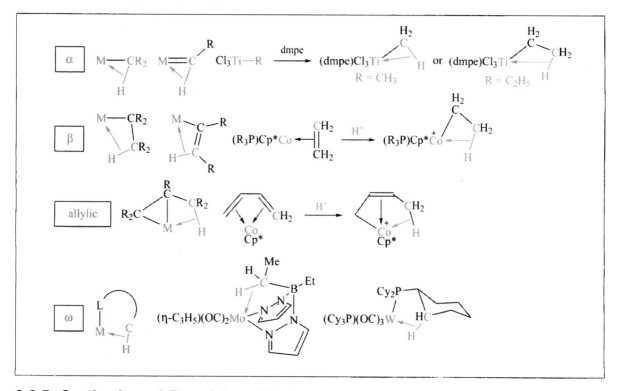

Figure 2.24 C–H–M agostic interactions

2.2.5 Synthesis and Reactivity of Hydride Complexes

The majority of synthetic approaches involve sources of H_2, H^+ or H^-. However, of particular interest will be the various ways in which organometallic ligands may decompose to provide hydride ligands, discussed in subsequent Chapters (*e.g.* α-elimination, β-elimination; see Chapters 4–6).

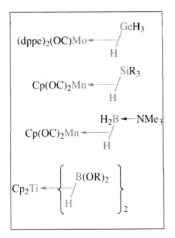

Figure 2.25 Heteroatom–H–M agostic interactions

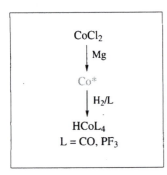

Figure 2.26 Hydride synthesis *via* Rieke-activated metals

Very few metals react directly with CO (Chapter 3); however, some will react with mixtures of CO and H_2, especially if the metal can be prepared in highly activated form, *e.g.* by reduction of metal halides *in situ*. More commonly, a metal halide or other salt is reduced in the presence of H_2/CO mixtures. In a similar manner, mixtures of H_2 and PF_3 can provide related PF_3 complexes (Figure 2.26).

The simplest method for the introduction of a hydride ligand involves the protonation of a metal centre (Figure 2.13). This is a conventional Brønsted acid–base equilibrium (see Table 2.4), dependent on the basicity of the metal centre. This is, in turn, a function of charge, oxidation state and co-ligands. Protonation does not alter the number of valence electrons, but leads to a formal increase in the oxidation state by two units and of the coordination number by one unit. Note also that protonation *may* be regioselective if alternative sites are available on the metal centre. This is illustrated in Figure 2.27a, which also demonstrates the use of an acid with a non-coordinating conjugate base (perchlorate). A range of other commonly employed 'non-coordinating' anions are shown in Figure 2.27b. However, it should be appreciated that some of these can in some cases coordinate, albeit weakly. In the cases of ClO_4^-, $CF_3SO_3^-$, BF_4^- and PF_6^- this is most easily established by IR spectroscopy, since coordination to a metal lowers the symmetry of the anion, leading to changes in the number and positions of (typically intense) Cl–O (see also Figure 3.18), S–O, B–F and P–F bands. When acids with potentially nucleophilic conjugate bases are employed, the possibility of coordination arises, thereby constituting oxidative addition. This is illustrated by the reactions of Vaska's complex $IrCl(CO)(PPh_3)_2$ with a range of 'protic' substrates (HX), some of which are normally considered to be quite weak acids (Figure 2.20). A special case of oxidative addition involves the reactions of metal complexes with hydrogen, in which the stereochemistry of the products is consistent with the intermediacy of dihydrogen complexes (see above). Similar observations for the additions of the Si–H bond of silanes suggest that prior coordination of the Si–H bond also occurs. Indeed, stable silane complexes have been isolated (Figure 2.25). A special case of H–X oxidative addition, which we will return to, involves the addition of C–H bonds (C–H activation). In the case of terminal alkynes and cyclopentadienes, these C–H bonds are sufficiently acidic for simple oxidative addition to occur. The osmium examples (Figure 2.28a) illustrate the effect of metal basicity on the facility of oxidative addition. Increasing the basicity of the phosphine co-ligands from PPh_3 to PPr^i_3 (see Table 2.1) favours the oxidative addition of the alkyne C–H. In Chapter 4 we will see that for some metal complexes even unfunctionalized alkanes may oxidatively add ('C–H activation', *e.g.* Figure 2.28c,d). This reaction is often reversible, with alkane reductive elimination being a facile decompostion route for *cis* metal alkyl hydrides.

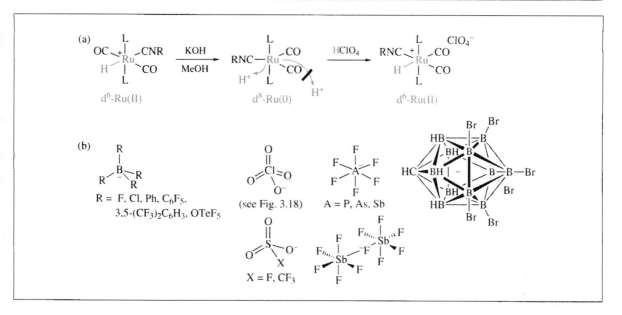

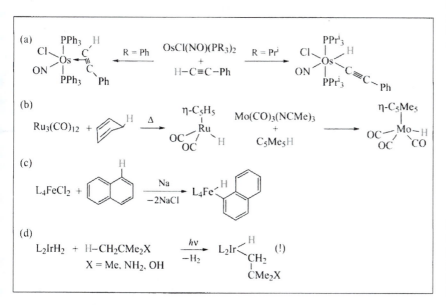

Figure 2.27 (a) Regioselective metal protonation (R = C₆H₄Me; L = PPh₃); (b) non- or weakly coordinating anions

Figure 2.28 C–H oxidative additions involving (a) alkynyl, (b) cyclopentadienyl, (c) aryl and (d) aliphatic C–H bonds; L₄Fe = Fe(dmpe)₂; L₂Ir = Ir(CO)Cp*, Ir(PMe₃)Cp*

The hydride ligand may also be introduced in a nucleophilic form through the reactions of metal halides or other salts with 'hydride' donors (Figure 2.29a). This may be thought of as a simple nucleophilic substitution of halo or solvento ligands by hydride. However, the nucleophile is not hydride itself but in most cases a hydride donor based on a Group 13 metal, among which Na[BH₄] and Li[AlH₄] and their derivatives (*e.g.* Li[HBEt₃], Li[HAl(OBuᵗ)₃]) are the most commonly employed. When alcohols are used as the solvent in these reactions, dihydrogen is also generated, leading in many cases to polyhydrides (*e.g.* Figure 2.17). In some cases, intermediate tetrahydroborato complexes may be isolated (Figure 2.29b). In the case of metal carbonyl substrates which are typically coordinatively saturated (18VE and therefore not electrophilic at the metal centre), formyl intermediates may be involved which thermally extrude a CO co-ligand, allowing the hydrogen to transfer from the formyl ligand to the metal centre (Figure 2.29c; Chapter 4).

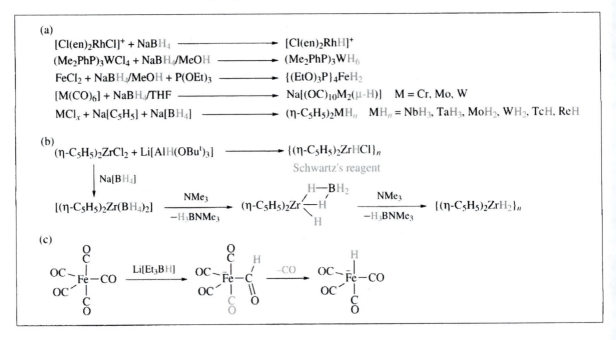

Figure 2.29 Metal hydride synthesis (a) employing borohydride reagents, (b) *via* tetrahydroborate ligands and (c) *via* formyl intermediates

2.2.6 Reactivity of Metal Hydrides

Transition metal hydrides may show hydridic ($H^{\delta-}$, Figure 2.16) or protic reactivity ($H^{\delta+}$, Table 2.4). In addition, there are some characteristic reactions that fall outside this categorization, which involve coordination of the substrate to the metal centre, prior to combination with the hydride ligand (Figure 2.30).

In an organometallic context, however, the migration of the hydride on to an organometallic *cis* co-ligand is particularly relevant and will be met often throughout this text (Figure 2.31). In many cases, these reactions are reversible; however, it should be noted that the forward reaction results in a depletion of 2VE. The resulting vacant coordination site may be filled by an incoming ligand, or alternatively the new ligand might also contain a donor group, thereby favouring the formation of the insertion product. We have already seen that H–H and Si–H σ-bonds may act as ligands. In a similar manner, migration of a hydride to an organic ligand may result in new ligands that have C–H σ-bonds which can weakly coordinate to the resulting vacant site in an agostic manner. This

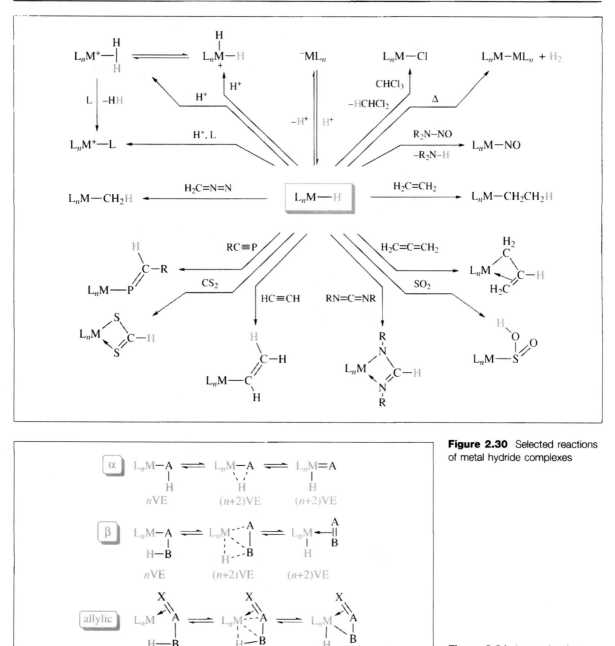

Figure 2.30 Selected reactions of metal hydride complexes

Figure 2.31 Intramolecular hydrogen transfers

is most obvious in crystallographically determined structures. However, when it persists in solution, it may be inferred on occasion from ^{1}H NMR data, owing to the characteristic upfield shift of agostically coordinated protons, and also in a decrease in the value of $^1J(CH)$. In the case of simple H–C(sp^3) bonds, this is typically *ca.* 125 Hz, but commonly falls below 100 Hz when agostic interactions occur with metals.

3
Carbon Monoxide

Aims

This chapter is concerned with carbon monoxide as a ligand. By the end of the chapter you should be familiar with:

- The structures and properties of the binary (homoleptic) carbonyl complexes of the transition elements, including representative synthetic routes
- How the structures which will be met reinforce the importance of the 18-electron rule
- The modified reactivity of CO when coordinated to a metal, and the variable ways in which the metal centre influences this reactivity

The discussion of typical reactions of coordinated CO should be seen as an introduction to the reactivity of other unsaturated molecules bound to, and activated by, transition metals.

3.1   Introduction

In 1888, Ludwig Mond reported that crude nickel powder reacted with CO under ambient conditions (the only metal to do so) to give a colourless volatile compound of composition C_4NiO_4. This observation was contemporary with Alfred Werner's fledgling ideas of primary and secondary valence in coordination compounds. Carbon monoxide was not, however, considered to show any basic or nucleophilic properties. Accordingly, the nature of this strange compound remained obscure, but this did not prevent an immediate application being devised: 'C_4NiO_4' could be thermally decomposed back into its constituents Ni and CO, providing a means of purifying crude nickel *via* a volatile and distillable

Ludwig Mond's nickel purification plant still operates to this day.

intermediate. This also suggested that the CO somehow retained its integrity in the curious compound.

In fact, metal carbonyl complexes occur in nature, detrimentally in the case of the strong adduct formed between haemoglobin and CO. Very recently, however, it has been found that the active site of hydrogenases isolated from *Desulfovibrio desulfuricans* and *Clostridium pasteurianum* feature diiron cores ligated with both CO and cyanide ligands (Figure 3.1a). Precise information about the structural features of the active sites of metalloproteins is often difficult to obtain crystallographically. However, as is often the case for chemical regimes not amenable to simple study, small molecule model compounds may yield to more complete chemical and spectroscopic characterization, (*e.g.* Figure 3.1b).

For a molecule with such a low basicity (only protonated in super-acidic media), CO is a surprisingly versatile ligand. The homoleptic (or binary) carbonyls (*i.e.* in which CO is the only ligand) already span oxidation states from –IV to +III, and higher oxidation states are known for heteroleptic complexes (mixed ligand sets) which have good π-donor co-ligands, *e.g.* $Ru^{IV}(CO)(SR)_4$.

3.2 Coordination Modes for Carbon Monoxide

By far the most common mode of CO coordination is terminal through carbon. The bonding for this mode of coordination has already been discussed (Figure 1.12). For bi- and polynuclear complexes (clusters), however, there exist further coordination possibilities in which the CO ligand bridges a metal–metal bond (Figure 3.2). Simple symmetric bridging (Figure 3.2a) is reminiscent of ketones (ν_{CO} IR absorptions are shifted to lower frequency, *cf.* organic carbonyls). The ligand is considered to provide 1VE to each metal, with M–C bond lengths typically longer than for terminal carbonyls. The CO ligand thereby exerts less steric pressure on the metal and bridging carbonyls are more frequently encountered for the smaller first-row 3d transition metals, *e.g.* $Fe_3(\mu\text{-}CO)_2(CO)_{10}$

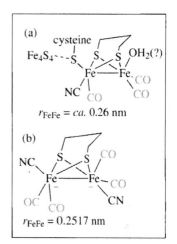

(a) cysteine

$r_{FeFe} = ca.\ 0.26$ nm

(b)

$r_{FeFe} = 0.2517$ nm

Figure 3.1 (a) A carbonyl complex from nature and (b) a synthetic model compound

Figure 3.2 Bridging carbonyl ligands

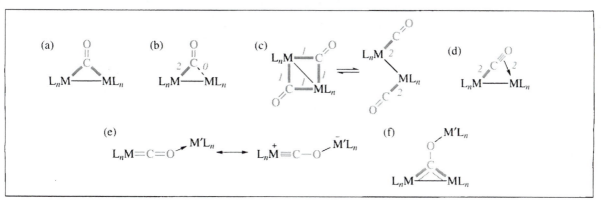

compared with $M_3(CO)_{12}$ (M = Ru, Os), which have only terminal ligands.

Semi-bridging carbonyls arise wherein the interaction with one metal (typically the more electron rich) is weaker (Figure 3.2b), although these probably represent the early stages of a continuum along the reaction coordinate depicted in Figure 3.2c. Very little energy is required for carbonyl ligands to move between terminal and bridging positions, especially if two carbonyl ligands can move synchronously [Figure 3.2c, see also $Co_2(CO)_8$ below]. For each of these bonding modes the CO provides only 2VE to the overall count. For more electron-deficient bimetallic complexes, a third possibility arises: the CO acts as a conventional teminal ligand (2VE) to one metal, but bends towards the second metal to allow the π-system to bind side on (η^2) and thereby contribute a further 2VE (Figure 3.2d). Figure 1.12 might suggest that the oxygen of coordinated CO has very little nucleophilic character. Situations do arise, however, where this may coordinate to very hard (early, electropositive) metal centres ('isocarbonyls', Figure 3.2e). Doubly or triply bridging CO ligands have more nucleophilic oxygen atoms such that coordination of a hard metal centre (or other electrophile) becomes more favourable (Figure 3.2f).

3.3 Characterization of Metal Carbonyls

For carbonyl complexes, a combination of IR (p. 11) and ^{13}C NMR spectroscopy will often reveal the molecular symmetry and also provide further information about the nature of the metal centre to which the CO ligand(s) is bound. However, the spectroscopic events involved occur within completely different time frames. Molecular vibrations (IR and Raman spectroscopy) are rapid relative to molecular fluxional processes. NMR transitions, however, are slow, often comparable in rate to intramolecular fluxionality and even intermolecular ligand exchange processes. This can lead to 'time-averaged' chemical environments being observed 'on the NMR timescale'. So long as this is borne in mind, ^{13}C NMR spectroscopy is a very valuable technique and can provide thermodynamic and kinetic data about such processes over a temperature range [variable temperature (VT) NMR].

Modern NMR spectrometers are routinely equipped with temperature-controlled probes.

This point is illustrated by the binary or homoleptic carbonyl complex $Co_2(CO)_8$ (Figure 3.3). In the solid state a structure with two bridging and six terminal carbonyl ligands is adopted. In hexane solution, however, a total of 13 v_{CO} bands are observed, a number far in excess of that expected for a single compound with only eight carbonyl ligands. This is due to three isomers being present in solution. One of these corresponds to the solid state isomer with bridging carbonyls [$v_{\mu CO}$ = 1866, 1857 cm^{-1}], but the remaining two have only terminal carbonyl ligands.

The ^{13}C NMR spectrum, however, comprises a single resonance (204 ppm), indicating that the three isomers interconvert very rapidly ($k >> 10^3 \text{ s}^{-1}$). This is a very common phenomenon in bi- and polynuclear (cluster) carbonyl complexes.

IR spectroscopy provides information about both the number and local symmetry of carbonyl ligands in a complex (Table 3.1). Note that the generalizations in this table assume that the remaining ligands are identical. If this is not the case, then the lower symmetry may lead to more absorptions. The medium for the measurement of IR spectra of metal carbonyls is important and typically spectra should be measured both in the solid state and in solution [*e.g.* $Co_2(CO)_8$]. Different structures may be adopted in the solid state and solution, a complex may crystallize in different crystal forms, or even have more than one crystallographic environment or molecule in the unit cell. The carbonyl ligands may therefore experience different environments and give rise to different v_{CO} values ('solid state splitting'). In solution (solvated), however, the orientations will be random. One important class of carbonyl complexes is the carbonyl metallates (see p. 48). For such salts the nature of the counter cation may be important in dictating both the frequency of the v_{CO} absorptions and the nucleophilicity of the metal, owing to the relative extent of ion pairing in different solvents, *e.g.* for $Na_2[Fe(CO)_4]$, $v(CO) = 1786$ (H_2O), 1730 (DMF) and 1761 (Nujol, *i.e.* solid state) cm^{-1}.

Table 3.1 Local $M(CO)_x$ symmetry consistent with the observed number of IR-active $v(CO)$ absorptions

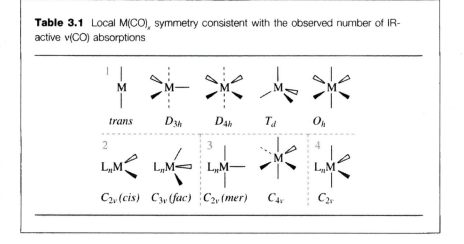

Terminal carbonyl ligands generally give rise to ^{13}C NMR resonances in the region 160–230 ppm, the chemical shift (δ) being most significantly dependent on the metal centre. In general, heavier metals cause shifts to higher field. Bridging carbonyls typically resonate to lower field.

3.4   Important Classes of Metal Carbonyls

3.4.1   Binary (Homoleptic) Carbonyls

The family of complexes that contain only CO are referred to as homoleptic or binary carbonyls and may be cationic, anionic or neutral. The more common neutral examples are illustrated in Figure 3.3, since these serve as key precursors for much metal carbonyl chemistry. Since CO is a strong π-acid, these involve metals in low (even formally negative) oxidation states, with low-spin electronic configurations (high Δ). With the exception of $V(CO)_6$ (t_{2g}^5), these are diamagnetic and generally coordinatively saturated, *i.e.* they obey the 18VE rule. Elements from odd groups (with the exception of vanadium) can only satisfy the 18-electron rule by forming either neutral bi- or polynuclear species or singly

CAUTION: neutral homoleptic metal carbonyls are volatile, lipophilic and highly toxic.

Figure 3.3 Selected binary metal carbonyls

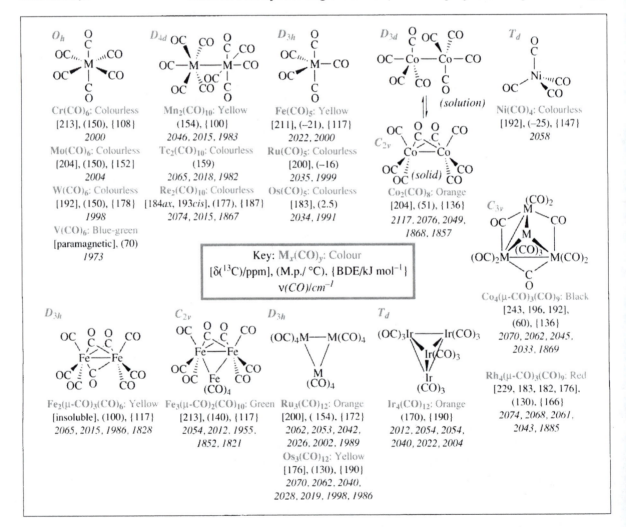

Key: $M_x(CO)_y$: Colour
$[\delta(^{13}C)/ppm]$, (M.p./ °C), {BDE/kJ mol^{-1}}
$\nu(CO)/cm^{-1}$

O_h
Cr(CO)$_6$: Colourless
[213], (150), {108}
2000
Mo(CO)$_6$: Colourless
[204], (150), {152}
2004
W(CO)$_6$: Colourless
[192], (150), {178}
1998
V(CO)$_6$: Blue-green
[paramagnetic], (70)
1973

D_{4d}
Mn$_2$(CO)$_{10}$: Yellow
(154), {100}
2046, 2015, 1983
Tc$_2$(CO)$_{10}$: Colourless
(159)
2065, 2018, 1982
Re$_2$(CO)$_{10}$: Colourless
[184ax, 193cis], (177), {187}
2074, 2015, 1867

D_{3h}
Fe(CO)$_5$: Yellow
[211], (–21), {117}
2022, 2000
Ru(CO)$_5$: Colourless
[200], (–16)
2035, 1999
Os(CO)$_5$: Colourless
[183], (2.5)
2034, 1991

D_{3d} (solution)
C_{2v} (solid)
Co$_2$(CO)$_8$: Orange
[204], (51), {136}
2117, 2076, 2049,
1868, 1857

T_d
Ni(CO)$_4$: Colourless
[192], (–25), {147}
2058

C_{3v}
Co$_4$(μ-CO)$_3$(CO)$_9$: Black
[243, 196, 192],
(60), {136}
2070, 2062, 2045,
2033, 1869

Rh$_4$(μ-CO)$_3$(CO)$_9$: Red
[229, 183, 182, 176],
(130), {166}
2074, 2068, 2061,
2043, 1885

D_{3h}
Fe$_2$(μ-CO)$_3$(CO)$_6$: Yellow
[insoluble], (100), {117}
2065, 2015, 1986, 1828

C_{2v}
Fe$_3$(μ-CO)$_2$(CO)$_{10}$: Green
[213], (140), {117}
2054, 2012, 1955,
1852, 1821

D_{3h}
Ru$_3$(CO)$_{12}$: Orange
[200], (154), {172}
2062, 2053, 2042,
2026, 2002, 1989
Os$_3$(CO)$_{12}$: Yellow
[176], (130), {190}
2070, 2062, 2040,
2028, 2019, 1998, 1986

T_d
Ir$_4$(CO)$_{12}$: Orange
(170), {190}
2012, 2054, 2054,
2040, 2022, 2004

charged mononuclear ions. In general, metal–metal bonding becomes increasingly favourable on descending a triad, such that polynuclear metal carbonyls are particularly common for the heavier (4d and 5d) elements.

Carbonyl metallates (anionic metal carbonyls) are an extremely important class of reagent in metal carbonyl chemistry, because they offer the synthetic advantage of a nucleophilic metal centre (see Figure 1.12, HOMO). Towards the far right-hand side of the transition series the metal π-basicity generally decreases as the p and d orbitals decrease in energy. Thus binary carbonyl complexes become more sparse and have increasingly labile carbonyl ligands. These may often be easily displaced by even comparatively non-nucleophilic ligands, including weakly nucleophilic solvents and traditionally 'non-coordinating' counter anions (Figure 2.27b). Synthetic routes typically involve the generation of ligand-free metal cations in highly Lewis acidic solvents with only non-coordinating anions being present. Thus CO is the only available potential ligand. In this respect SbF_5 (or 'super-acidic' $SbF_5/HOSO_3F$) is a particularly effective solvent for the isolation of $[SbF_6]^-$ or $[Sb_2F_{11}]^-$ salts of metal carbonyl complexes. Even under these conditions, the carbonyl ligands tend to be labile (very high v_{CO}) and many of the salts decompose readily. In such syntheses (Figure 3.4), either the elemental metal is oxidized *in situ*, or salts of an appropriate anion (*e.g.* SO_3F^-) may be employed. Under such conditions, CO may be readily oxidized, *e.g.* in the reductive carbonylation of $Pt(O_3SF)_4$. Table 3.2 summarizes the important isolable mononuclear homoleptic carbonyl complexes.

In contrast to the highly reduced early transition metallates (Table 3.2), the CO binding in the cationic complexes shown in Figure 3.4

Table 3.2 Homoleptic mononuclear carbonyl complexes[a] $[M(CO)_x]^y$

x	y	M	x	y	M
1	+1	Cu	5	−3	V, Nb, Ta
2	+1	Cu, Ag, Au	5	−2	Cr, Mo, W
2	+2	*Hg*	5	−1	Mn, Tc, Rh
3	−3	Co, Rh, Ir	5	0	Fe, Ru, Os
4	−4	Cr, Mo, W	6	−2	Ti, Zr, Hf
4	−3	Mn, Re	6	−1	V, Nb, Ta
4	−2	*Ti, Zr, Hf*	6	0	V, Cr, Mo, W
		Fe, Ru, Os	6	+1	Mn, Tc, Re
4	−1	Co, Rh, Ir	6	+2	Fe, Ru, Os
4	0	Ni	6	+3	Ir
4	+2	*Pd, Pt*			

[a]Complexes in italics have <18VE

			$\nu(CO)/cm^{-1}$	$\delta(^{13}C)/ppm$
		$[H{-}CO]^+$	2184	
$Au(O_3SF)_3 + CO + SbF_5$	$\longrightarrow$	$[Au(CO)_2][Sb_2F_{11}]$	2196	174
$Hg(O_3SF)_2 + CO + SbF_5$	$\longrightarrow$	$[Hg(CO)_2][Sb_2F_{11}]_2$	2278	169
$Pt(O_3SF)_4 + CO + SbF_5$	$\longrightarrow$	$[Pt(CO)_4][Sb_2F_{11}]_2$	2244	137
$IrF_6 + CO + SbF_5$	$\longrightarrow$	$[Ir(CO)_6][Sb_2F_{11}]_3$	2254	121
$Fe(CO)_5 + CO + AsF_5 + SbF_5$	$\longrightarrow$	$[Fe(CO)_6][Sb_2F_{11}]_2$	2203	179

Figure 3.4 Homoleptic carbonyl cations of the late transition metals

primarily involves σ-donation, with retrodonation being negligible. Accordingly, the values for ν_{CO} are actually higher than those for free or protonated CO. ^{13}C NMR chemical shifts for the carbonyl ligands of these late transition metal complexes occur to much higher field, *e.g.* in the isoelectronic d^6 complexes $[M(CO)_6]^n$: $M^n =$ Hf^{2-} 244, Ta$^-$ 210, W^0 192, Re$^+$ 171, Os^{2+} 147, Ir^{3+} 121 ppm. A similar situation arises for d^0 metal carbonyls {*e.g.* $[Ti(CO)_2Cp_2]^{2+}$, $\nu_{CO} = 2119$ and 2099 cm^{-1}; $[Zr(CO)(\eta^2\text{-}OCMe)Cp^*_2]^+$, $\nu_{CO} = 2176$ cm^{-1}}, where retrodonation is not possible. Cationic homoleptic isocyanide complexes of the late transition metals are more common, *e.g.* $[Rh(CNBu^t)_4]^+$ and $[Pt(CNMe)_4]^{2+}$. Although isocyanides are isoelectronic with CO, they are stronger σ-donors and weaker π-acceptors (Figure 1.14). Cyanide is also isoelectronic with CO, but a much stronger σ-donor. Hence homoleptic cyanides of the later transition metals are commonplace and some were known in the 18th century, *e.g.* $K_4[Fe(CN)_6]$.

3.4.2 Carbonyl Metallates

The ability of carbon monoxide to stabilize low oxidation states is most impressively demonstrated by the synthesis and isolation of complexes wherein the metal adopts formally negative oxidation states, of which the d^{10} complexes $[M(CO)_4]^{4-}$ (M^{-IV} = Cr, Mo, W) represent the (current) extreme.

By far the most common method for the synthesis of carbonyl metallates is *via* the reduction of binary carbonyls (or carbonyl halides). A variety of reducing agents has been employed including alkali metal amalgams, sodium in liquid ammonia, borohydride derivatives, *e.g.* Li[Et$_3$BH], and alkali metals in combination with (often catalytic) polyaromatic organic molecules capable of serving as electron transfer reagents (*e.g.* benzophenone, anthracene, naphthalene). In the case of mononuclear precursors, two-electron reduction (+2VE) is accompanied by loss of one carbonyl ligand (–2VE) to satisfy the 18-electron rule (Figure 3.5a). Occasionally, it is advantageous to begin with a derivative which contains a labile purely σ-bonding ligand which will more easily dissociate from the reduced metal centre [*e.g.* W(NMe$_3$)(CO)$_5$, Figure 3.5a], thereby avoiding side reactions, *e.g.* redox condensation (see below). For bi- and polynuclear precursors, reduction is accompanied by metal–metal bond cleavage (Figure 3.5b). Stronger reductants (Na[C$_{10}$H$_8$] or Na[Ph$_2$CO]) may be employed to provide more highly reduced species ([M(CO)$_4$]$^{n-}$, M^{n-} = Mn^{3-}, Cr^{4-}). Group 13 hydride reagents may also be employed to cleave metal–metal bonds, in which case metal hydride complexes are implicated (or isolated) as

Figure 3.5 Synthesis of carbonyl metallates

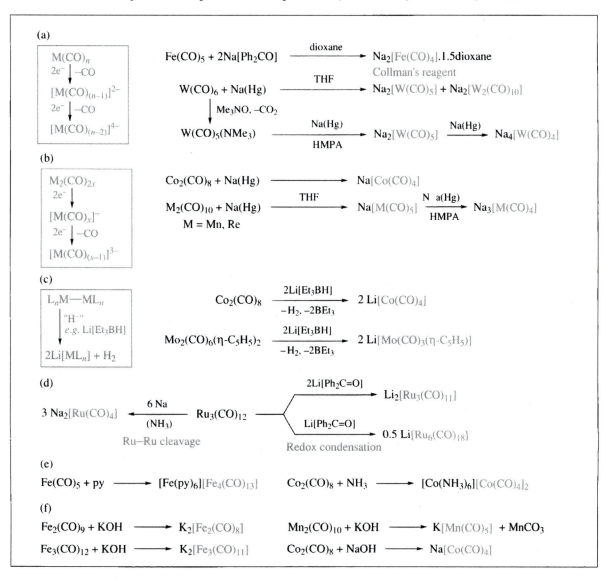

intermediates (Figure 3.5c). A further route to metal carbonylates involves the redox disproptionation of metal carbonyls, which may be induced by strong nucleophiles (Figure 3.5e). Although this is a key reaction of metal carbonyls, it is synthetically unattractive so far as metal atom economy is concerned. Sources of hydroxide can also bring about reduction of metal carbonyls; however, this is mechanistically more complex since coordinated CO is oxidized to CO_2 (or carbonate) during the process (Figures 2.13, 3.5f).

Carbonyl metallates find their widest application as reagents for introducing electrophilic functionality to the metal centre, which is highly nucleophilic. Table 3.3 indicates nucleophilicities for a range of common carbonyl metallates estimated from conventional S_N2 reaction rates with iodomethane. This illustrates their versatility in metal–carbon bond forming reactions; however, as shown in Figure 3.6, this reactivity is not limited to carbon electrophiles but allows a wide range of metal–element bonds to be easily formed (see also reactions of metal carbonyls below).

Alkali metal salts of carbonyl metallates are often **pyrophoric** (spontaneously flammable in air); however, in many cases they may be isolated more conveniently as salts of large organic cations, e.g. [N(PPh₃)₂]⁺ (PPN⁺) and [AsPh₄]⁺

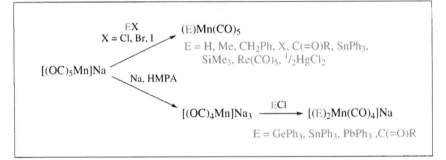

Figure 3.6 Reactions of carbonyl manganates with electrophiles

Two potential problems (opportunities?) which may be encountered are (i) redox condensation, whereby carbonylates of higher nuclearity may be formed (Figure 3.5a,d,e); effectively, the reduced carbonylate replaces a carbonyl ligand on combination with a second equivalent of the original neutral carbonyl; (ii) over-reduction, whereby more than two electrons may be added, *e.g.* in the formation of $Na_4[Cr(CO)_4]$ or $Na_3[Mn(CO)_4]$ (Figure 3.5b). These problems are controlled by judicious (or empirical) choice of solvent and reductant.

3.4.3 Cyclopentadienyl Metal Carbonyls

Chemistry involving transformations of the cyclopentadienyl (Cp) ligand itself will be discussed in Chapter 7; however, much of the chemistry of cyclopentadienyl complexes involves the ligand acting purely as a spectator. Thus there arises a broad area of chemistry involving complexes with both carbonyl and cyclopentadienyl ligands. The most common classes of such compounds (Table 3.4) are the mononuclear complexes

Table 3.3 Relative nucleophilicities of carbonyl metallates

Anion	Relative rates[a]
[Fe(CO)₂Cp]⁻	7.0×10^7
[Ru(CO)₂Cp]⁻	7.5×10^6
[Ni(CO)Cp]⁻	5.5×10^6
[Re(CO)₅]⁻	2.5×10^4
[W(CO)₃Cp]⁻	5×10^2
[Mn(CO)₅]⁻	77
[Mo(CO)₃Cp]⁻	67
[Cr(CO)₃Cp]⁻	4
[Co(CO)₄]⁻	1

[a]For reactions with MeI

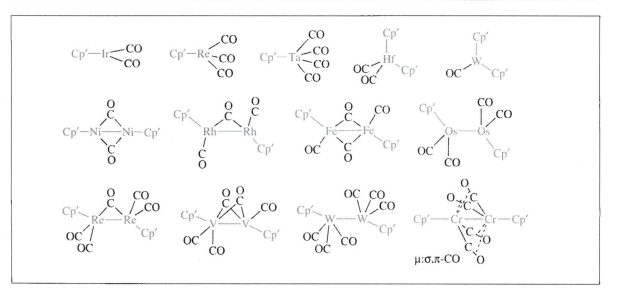

μ:σ,π-CO

Figure 3.7 Representative cyclopentadienyl metal carbonyls (Cp′ = η-C₅R₅)

$M(CO)_y(\eta\text{-}C_5R_5)_z$ and the binuclear species $M_2(CO)_y(\eta\text{-}C_5R_5)_z$, the values of y and z being largely dictated by adherence to the 18VE rule. Figure 3.7 illustrates structural motifs for the more commonly encountered neutral examples, whilst Figure 3.8 illustrates the most common synthetic routes. Noteworthy amongst these is the reaction of mononuclear metal carbonyls with cyclopentadiene (C_5H_6) to provide binuclear derivatives. Hydrogen is eliminated, and the most likely intermediates involve cyclopentadienyl metal hydrides that thermally eliminate dihydrogen in a bimolecular process (Figure 2.28). The binuclear derivatives are key starting materials for much chemistry, owing to the often facile reduction of the metal–metal bond to provide mononuclear nucleophilic carbonyl metallates, e.g. $[Cr(CO)_3Cp]^-$, $[Fe(CO)_2Cp]^-$ and $[Ni(CO)Cp]^-$.

In some cases, complexes arise which may appear to have more than the requisite 18VE and, in such situations, one of the cyclopentadienyl rings may assume reduced hapticity (η^3 or even η^1, Figure 3.9). This is more pronounced in complexes of the indenyl ligand (C_9H_7) where 'ring-slippage' may be compensated for by the increase in aromatic character within the benzene ring. Indeed, indenyl complexes are generally found to be more reactive towards ligand substitution processes than simple cyclopentadienyl analogues. These reactions appear to proceed via associative activation, suggesting that a vacant coordination site may be transiently achieved by ring slippage (known as the indenyl effect). In other cases, the metal centres in bi- or polynuclear complexes may appear to have <18VE, in which case the coordinative unsaturation is accommodated through metal–metal multiple bonding, or the adoption of σ–π coordination (Figure 3.2d) by one or more carbonyl ligands which thereby each contribute 4VE in total. Thus the simple stoichiometries

Table 3.4
Cyclopentadienyl metal carbonyls

$Cp_xM_y(CO)_z$[a]	M
$Cp'M(CO)_2$	Co, Rh, Ir
$Cp'M(CO)_3$	Mn, Tc, Re
$Cp'M(CO)_4$	V, Nb, Ta
$Cp'_2M(CO)$	Mo, W, V
$Cp'_2M(CO)_2$	Ti, Zr, Hf
$Cp'_2M_2(CO)_2$	Ni, Pt
	Co, Rh, Ir
$Cp'_2M_2(CO)_3$	Co, Rh, Ir, Re
$Cp'_2M_2(CO)_4$	Fe, Ru, Os
	Cr, Mo, W
$Cp'_2M_2(CO)_5$	V, Re
$Cp'_2M_2(CO)_6$	Cr, Mo, W
$Cp'_3M_3(CO)_2$	Ni
$Cp'_3M_3(CO)_3$	Co, Rh
$Cp'_3M_3(CO)_7$	Nb
$Cp'_4M_4(CO)_4$	Fe

[a] $Cp' = \eta\text{-}C_5R_5$

Figure 3.8 Synthesis of cyclopentadienyl metal carbonyls

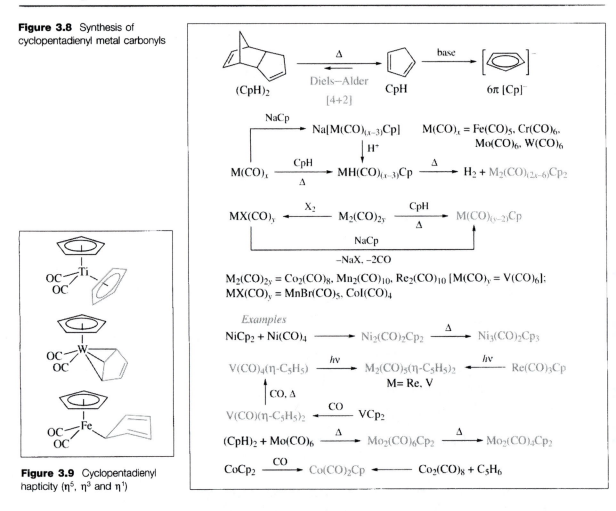

Figure 3.9 Cyclopentadienyl hapticity (η^5, η^3 and η^1)

indicated in Table 3.4 obscure more subtle bonding situations (or fluxional processes). As for simple binary carbonyls, thermolysis or photolysis generally leads to loss of CO and formation of higher nuclearity clusters.

3.4.4 Nitrosyl Metal Carbonyls

The nitrosonium cation $[NO]^+$ is isoelectronic with CO and accordingly many mixed nitrosyl-carbonyl complexes are known. For electron counting purposes, the neutral molecule is considered to act as a 3 (or occasionally 1) VE donor. Thus various series of isoelectronic complexes can be envisaged (Table 3.5). The majority of synthetic routes to nitrosyl-carbonyl complexes involve (i) photochemical CO substitution or metal–metal bond cleavage by NO; (ii) electrophilic attack by nitrosonium salts, *e.g.* $[NO]BF_4$ or nitrosyl halides (*e.g.* ClNO) upon electron-rich metal complexes; (iii) oxidation of coordinated CO by nitrite accompanied by reduction of the metal centre; or (iv) reactions of metal hydrides with nitrosamines or nitrite esters (Figure 3.10). In this context we consider the nitrosyl ligand to serve as a 3VE ligand; however, note that it may also act as a 1VE ligand by bending (reversibly) at nitrogen. Thus it is commonly observed that nitrosyl complexes can undergo ligand substitution processes *via* associative activation, *i.e.* an increase in coordination number, without contravention of the 18-electron rule. For example, whilst Fe(CO)$_5$ undergoes CO substitution *via* a dissociative process, the reactions of Mn(NO)(CO)$_4$ with phosphines show a dependence on the nucleophilicity of the incoming ligand (*e.g.* PBu$_3$ reacts 40 times faster than PPh$_3$). This suggests a bimolecular rate-determining step involving an increase in coordination number which may be accommodated by nitrosyl bending.

Table 3.5 Isoelectronic carbonyl nitrosyl complexes[a]
d^{10} ML$_4$
Ni(CO)$_4$
Co(NO)(CO)$_3$
Fe(NO)$_2$(CO)$_2$
Mn(NO)$_3$(CO)
Cr(NO)$_4$
d^8 ML$_5$
Fe(CO)$_5$
Mn(NO)(CO)$_4$
[Mn(CO)$_5$]$^-$
[Cr(NO)(CO)$_4$]$^-$
d^6 CpML$_3$
[CpFe(CO)$_3$]$^+$
[CpMn(NO)(CO)$_2$]$^+$
[CpCr(NO)$_2$(CO)]$^+$
[CpV(NO)$_3$]$^+$
CpMn(CO)$_3$
CpCr(NO)(CO)$_2$
CpV(NO)$_2$(CO)
[CpCr(CO)$_3$]$^-$

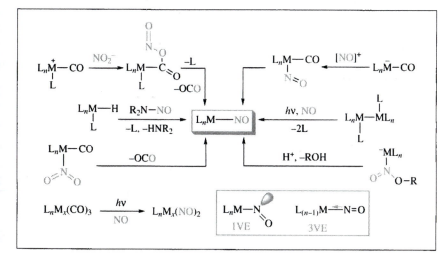

CAUTION: nitrosamines and nitrite esters are potent carcinogens.

Figure 3.10 Synthesis of nitrosyl metal carbonyls

3.5 Synthesis of Metal Carbonyls

The majority of synthetic approaches to binary carbonyls involve carbonylation of solutions of metal salts in the presence of a suitable reductant (Figure 3.11). Nickel and iron are the only metals that will react readily with CO, and even iron requires the use of high temperatures (420 K) and pressures (100 atm). Some other metals (Mo, W, Co, Ru) in highly activated form ('Rieke metals') will also react with CO under higher temperatures and pressures.

Both its solubility and any potential side reaction with reductant dictate the choice of precursor metal salt or complex. In many cases the reductant is an aluminium alkyl or electropositive metal (Goups 1, 2, 12 or 13, possibly amalgamated with mercury), requiring the use of anhydrous metal salts or complexes. Carbonyls of higher nuclearity are

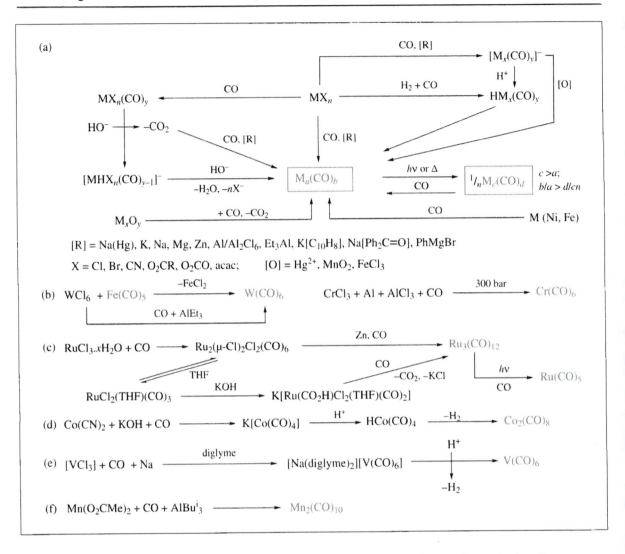

(a)

[R] = Na(Hg), K, Na, Mg, Zn, Al/Al$_2$Cl$_6$, Et$_3$Al, K[C$_{10}$H$_8$], Na[Ph$_2$C=O], PhMgBr

X = Cl, Br, CN, O$_2$CR, O$_2$CO, acac;   [O] = Hg^{2+}, MnO$_2$, FeCl$_3$

(b) WCl$_6$ + Fe(CO)$_5$ $\xrightarrow{-FeCl_2}$ W(CO)$_6$ CrCl$_3$ + Al + AlCl$_3$ + CO $\xrightarrow{300\ bar}$ Cr(CO)$_6$

(c) RuCl$_3$.xH$_2$O + CO $\longrightarrow$ Ru$_2$(μ-Cl)$_2$Cl$_2$(CO)$_6$

(d) Co(CN)$_2$ + KOH + CO $\longrightarrow$ K[Co(CO)$_4$] $\xrightarrow{H^+}$ HCo(CO)$_4$ $\xrightarrow{-H_2}$ Co$_2$(CO)$_8$

(e) [VCl$_3$] + CO + Na $\xrightarrow{diglyme}$ [Na(diglyme)$_2$][V(CO)$_6$] $\xrightarrow[-H_2]{H^+}$ V(CO)$_6$

(f) Mn(O$_2$CMe)$_2$ + CO + AlBu$^i_3$ $\longrightarrow$ Mn$_2$(CO)$_{10}$

Figure 3.11 (a) Generalized and (b)–(f) illustrative syntheses of binary carbonyls

generally prepared by either photolysis or thermolysis of mono- or binuclear precursors, although these may not necessarily be observed (Figure 3.12). The reverse of this process may be used to degrade higher nuclearity clusters by heating under high pressures of CO ('cluster decapping' reactions; Figure 3.12). Finally, the combination of CO and hydroxide offers a reducing medium *via* a process to be discussed below (Figures 2.13, 3.11).

Heteroleptic (mixed ligand) carbonyl complexes are intriguing in their enormous variability. Many of these derivatives originate *via* ligand substitution reactions starting from the homoleptic carbonyls (Figure 3.13). The replacement of strongly π-acidic CO by more basic ligands will electronically enrich the metal centre, strengthening retrodative bonding to any remaining CO ligands. However, if the new ligand (*e.g.* PF$_3$) is also a strong π-acid, the remaining CO ligands may still be labile (Figure 3.13a). Since binary carbonyls are coordinatively saturated, a vacant site must first be generated (dissociative activation, reduced coordination number). This may be achieved thermally, photochemically or by chemical modification of a carbonyl ligand. Given a choice between which CO ligands are substituted, it is usually one of a pair that is mutually *trans* coordinated (minimize competition for metal retrodonation). For example, heating $Mo(CO)_6$ with piperidine (pip) or acetonitrile provides *cis*-$Mo(CO)_4(pip)_2$ and *fac*-$Mo(CO)_3(NCMe)_3$, respectively, rather than the corresponding *trans* and *mer* isomers. These complexes are useful precursors for other derivatives of the '$Mo(CO)_4$' and '$Mo(CO)_3$' fragments owing to the facile substitution of the amine or nitrile ligands under much milder conditions (typically ambient) than those required for the initial thermal CO substitution.

Thermal reactions can be somewhat indiscriminate, and lead to mixtures of substitution products, *e.g.* Figure 3.13c,d. Photochemical substitution is often more selective, but less conveniently applicable to large-scale syntheses. For example, photolysis of the $(t_{2g})^6(e_g{}^*)^0$ complex $Cr(CO)_6$ will promote electrons into the previously empty $e_g{}^*$ orbitals, which are antibonding with respect to the Cr–CO bonds. Thus the photo-excited molecule can relax by ejection of one CO ligand. Typically this is carried out in a weakly coordinating solvent, *e.g.* tetrahydrofuran (continuously purged with nitrogen to remove liberated CO) to provide a labile deep red *solvento* complex $Cr(CO)_5(THF)$. Once photolysis is

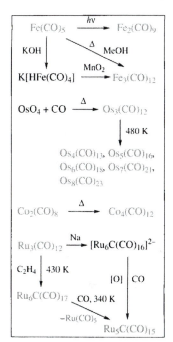

Figure 3.12 Metal carbonyl interconversions involving changes in nuclearity

Figure 3.13 Thermal and photochemical carbonyl substitution

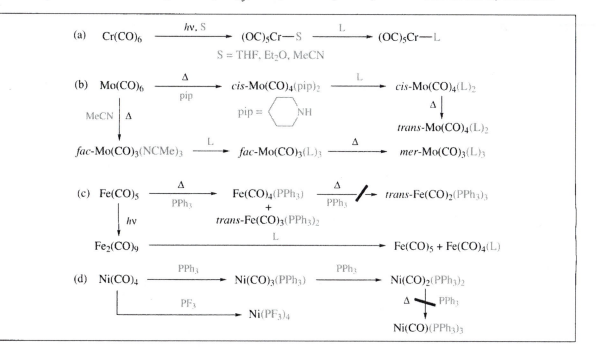

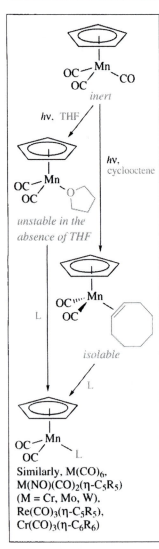

Similarly, M(CO)$_6$,
M(NO)(CO)$_2$(η-C$_5$R$_5$)
(M = Cr, Mo, W),
Re(CO)$_3$(η-C$_5$R$_5$),
Cr(CO)$_3$(η-C$_6$R$_6$)

Figure 3.14 Photochemical generation of *solvento* intermediates

complete (typically monitored by solution IR spectroscopy), this may then be used *in situ* for subsequent facile substitution of the THF ligand by more strongly coordinating ligands, including those which might be too photo or thermally sensitive to survive direct thermolysis or photolysis (Figure 3.14). Occasionally, alkenes (Chapter 6) may be used as a compromise between the inert CO coordination and the extremely labile solvent coordination, *e.g.* the photochemically generated complexes Mn(η^2-C$_8$H$_{14}$)(CO)$_2$Cp and Cr(η^2-C$_8$H$_{14}$)(CO)$_5$ (C$_8$H$_{14}$ = cyclooctene) serve as isolable (by chromatography) alternatives to the highly reactive complexes Mn(THF)(CO)$_2$Cp and Cr(THF)(CO)$_5$ (unstable in the absence of THF) as sources of the 16VE fragments 'Mn(CO)$_2$Cp' and 'Cr(CO)$_5$'.

Carbonyl ligands may also be removed by treatment with an amine oxide; this oxidizes coordinated CO to CO$_2$, which does not coordinate strongly and is displaced by the (labile) amine (Figures 3.15 and 3.5a). Since this involves nucleophilic attack at coordinated CO, it is most effective for carbonyl ligands bound to less π-basic metal centres (higher v_{CO}).

An alternative means of activating coordinatively saturated metal carbonyls involves electron transfer catalysis. A strong one-electron reductant (*e.g.* Na[C$_{10}$H$_8$] or Na[Ph$_2$C=O]) adds an electron to the complex, converting it to a labile 19VE species. Replacement of CO by a stronger net donor makes the product a stronger reductant than the precursor and accordingly it is able to transfer an electron to a further molecule of substrate; hence catalysis can ensue (Figure 3.16).

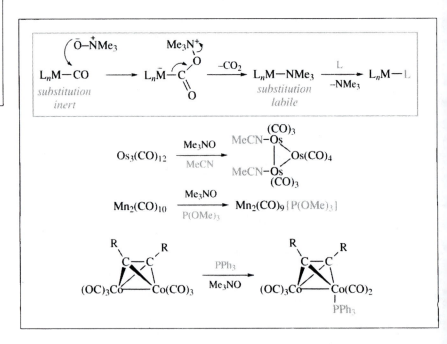

Figure 3.15 Amine oxide-mediated carbonyl substitution

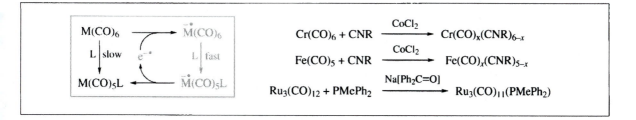

The alternative approach to heteroleptic carbonyl complexes involves the addition of CO to a preformed complex which either has a vacant coordination site to begin with, or is capable in solution of eliminating one or two ligands, *e.g. via* simple ligand dissociation, reductive elimination or ligand coupling processes (Figure 3.17). In the case of halide complexes, salts of non-coordinating anions (Figure 2.27b) may be used to extract the halide, leaving a vacant coordination site (Figure 3.18). Such reactions are equilibria, often driven by the precipitation of the halide salt. This is particularly favourable for thallium(I) or silver(I) salts, where the resulting halide is essentially insoluble. Silver salts can, however, also act as one-electron oxidants (formation of Ag^0), in which case the highly toxic though non-oxidizing thallium(I) salts may be used instead. Various Lewis acids are also capable of abstracting halide to generate a non-coordinating anion, *e.g.* Al_2Cl_6 (see also Figure 3.4)

Figure 3.16 Electron transfer-catalysed carbonyl substitution

Figure 3.17 Methods for the generation of a vacant coordination site (acac = propane-2,4-dionate, acetylacetonate)

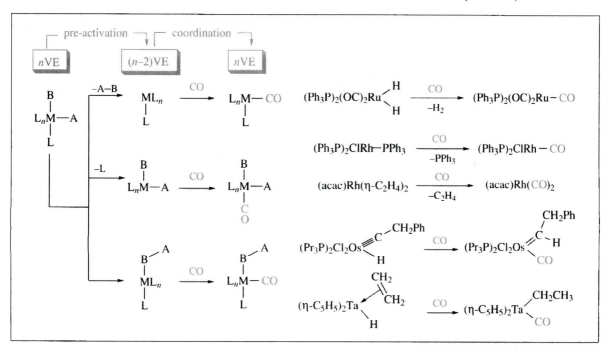

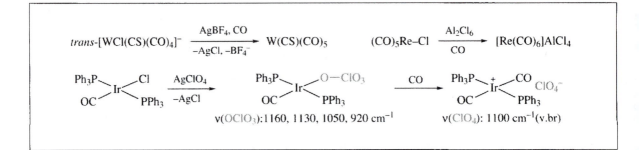

Figure 3.18 Halide abstraction/CO coordination

3.6   Reactions of Metal Carbonyls

It is useful to consider the reactions of carbonyl metallates separately, since their reactivity is generally concerned with the nucleophilic metal centre and will be discussed below. Simple ligand substitution reactions have already been discussed above, as have redox processes that provide access to carbonyl metallates through reduction of the metal centre. These redox or ligand addition/elimination processes are in principle no different from those encountered for classical ligands. We will now consider reactions in which the carbonyl ligand itself enters directly into the reaction and emerges transformed.

3.6.1 Electrophilic Attack

Figure 1.12 suggests that for carbonyl complexes the HOMO is localized primarily on the metal centre, with only a modest contribution from oxygen orbitals. Thus by far the majority of reactions of metal carbonyls with electrophiles involve direct attack at the metal, with the carbonyl serving as a spectator ligand. If, however, the metal centre is (i) particularly electron rich and (ii) sterically shielded and the electrophile is 'hard' (in the HSAB sense) and also sterically encumbered, then attack may occur at the oxygen. Thiocarbonyls (L_nM–CS) are stronger π-acids than CO and the sulfur is both softer and more nucleophilic. Thus electrophilic attack at the sulfur of thiocarbonyls is more common if the metal centre is electron rich ($v_{CS} < 1200$ cm^{-1}). Similarly, coordinated isocyanides (CNR) are more prone to attack by electrophiles at nitrogen. This is noteworthy in the sense that free isocyanides are attacked by electrophiles at carbon (Figure 3.19). The resulting carbyne ligands will be discussed in Chapter 5.

3.6.2 Nucleophilic Attack

Although free CO is attacked by very strong nucleophiles (*e.g.* Bu^{t}Li), the resulting acyl anions are generally unstable. In contrast, coordinat-

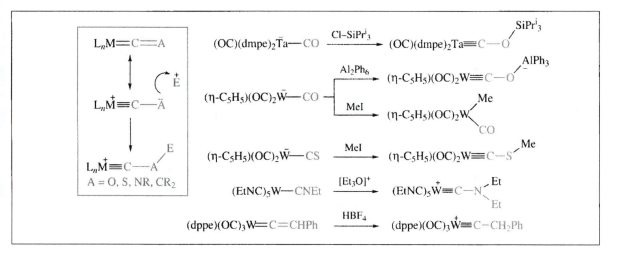

ed CO is highly prone to attack by nucleophiles, especially when the metal centre is not particularly π-basic (high v_{CO}) (Figure 3.20). Nucleophiles are potential ligands (and *vice versa*); if the metal centre is coordinatively unsaturated, attack may occur at the metal (although this may be followed by migration to CO; see below). For coordinatively saturated metal carbonyls, however, nucleophilic attack generally occurs at the carbonyl carbon, consistent with the high contribution of C orbitals to the LUMO (Figure 1.12).

Figure 3.19 Electrophilic attack at coordinated CO and related ligands

Figure 3.20 Nucleophilic attack at coordinated CO and related ligands

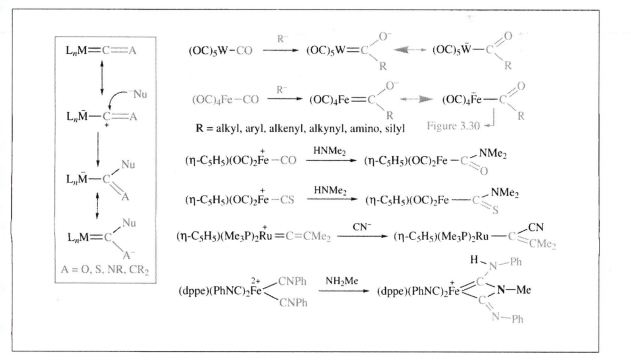

The resulting acyl metallates may be described by the two canonical forms shown, with a negative charge on either the metal or the oxygen. The relative contributions of these two forms will depend on the nature of the metal centre (d configuration, charge, co-ligands, coordination number; see also Figure 1.17). Therefore although they may be prone to subsequent attack by electrophiles, the regiochemistry of attack (metal *vs.* oxygen) may differ between complexes; thus the six-coordinate d^6 tungsten acylate is attacked by $[Me_3O]BF_4$ (Meerwein's reagent; synthetically equivalent to 'Me$^+$' and Me_2O) at the acyl oxygen to provide a carbene complex (Chapter 5). The five-coordinate d^8 iron example, however, undergoes attack at the metal centre followed by reductive elimination of a ketone (see Collman's reagent, Figures 3.5 and 3.29). Three special cases should be considered (Figure 3.21): hydride (Figure 2.29c), hydroxide (Figure 2.13) and nitrite (Figure 3.10).

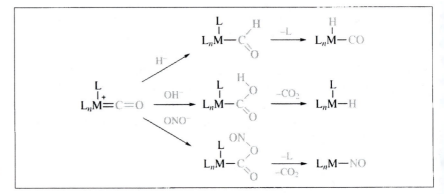

Figure 3.21 Nucleophilic attack at coordinated CO: special cases

In general, the initially formed formyl, hydroxycarbonyl or nitrito-carbonyl complexes rapidly convert to hydrido or nitrosyl complexes with elimination of CO or CO_2; however, occasionally intermediates have been isolated. The oxidation of coordinated CO to CO_2 by amine oxides (Figure 3.15) may be considered a further example. The hydroxycarbonyl example underpins a technologically important process, the water-gas shift equilibrium, involving the catalytic conversion of CO and water into CO_2 and hydrogen (Figures 3.22, 3.11 and 2.13).

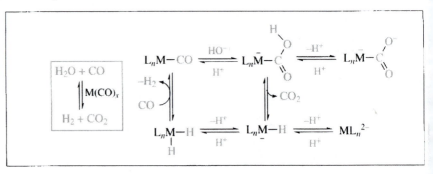

Figure 3.22 The water-gas shift equilibrium

3.6.3 Ligand–CO Coupling Processes: Migratory Insertion

If the π^* orbital of coordinated CO is sufficiently electrophilic to interact with external nucleophiles, the question arises as to whether the electron pair binding an adjacent ligand to the metal might not also act as an internal electrophile. This is indeed what often happens, especially if the adjacent ligand is a σ-organyl. Three cases deserve attention (the first being by far the most important): when the organic ligand is singly, doubly or triply bound to the metal (Figure 3.23). The situations involving multiply bound ligands (carbenes and carbynes) will be discussed in Chapter 5.

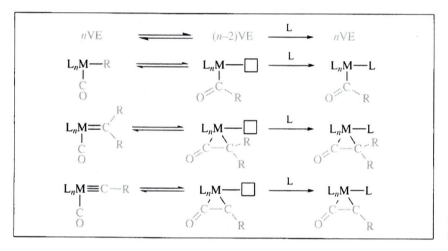

Figure 3.23 Carbonyl co-ligand coupling reactions

The case of simple σ-organyl ligands (alkyls, aryls, *etc.*) coupling with carbon monoxide is known as migratory insertion (Figure 3.24). The forward process should be viewed as the migration of the R group onto the carbonyl, thereby leaving its original coordination site vacant, and depleting the electron count by 2VE. This loss of coordinative saturation may appear to be an unfavourable development. Indeed, it is seldom directly observed and coordinatively unsaturated acyl complexes are rare for metals where the 18-electron rule holds sway. Three situations do, however, arise where the process may be inferred: (i) when a new ligand may be added to block the vacated coordination site; (ii) when the metal is sufficiently oxophilic for the oxygen of the resulting acyl to serve as an electron pair donor to the metal; and (iii) when subsequent reductive elimination of the acyl ligand may displace the natural position of the equilibrium. It is this latter situation that underpins the majority of technological applications of the process (*e.g.* see the Monsanto process, below). Illustrative examples are provided in Figure 3.25.

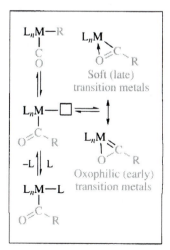

Figure 3.24 Equilibria manifold for migratory insertion

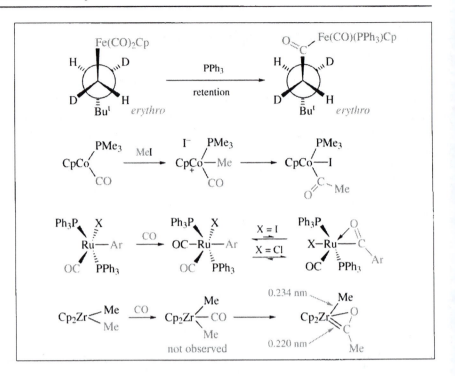

The mechanism of migratory insertion has been studied in considerable detail and some conclusions are summarized in Box 3.1. Much information has been extracted from studies on the system RMn(CO)$_5$/RCOMn(CO)$_5$, because of the inherent simplicity which allows careful control of the different variables (Figure 3.26).

Figure 3.26 Migratory
insertion reactions of Mn(Me)(CO)$_5$
($\mathbb{C}$ = ^{13}C)

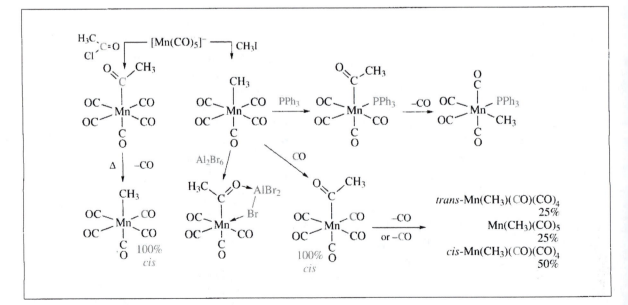

Box 3.1 Migratory Insertion: Evidence and Conclusions

- The introduced ligand L assumes a position *cis* to the resulting acyl ligand.
- When the group R is chiral at the migrating carbon, the chirality is retained (excludes intermolecular processes *via* ionic or radical intermediates which would readily invert/racemize).
- The process is accelerated by Lewis acids capable of co-ordinating to the carbonyl oxygen, thereby increasing the electrophilicity of the carbon.
- Isotopically labelled acyls $MC(=O)R$ de-insert (α-eliminate) to place the labelled CO *cis* to the resulting M–R group.
- The reaction may be catalysed by one-electron oxidants, *e.g.* $[FeCp_2]PF_6$ (see Chapter 7), which in addition to disrupting the 18VE configuration, also increase the electrophilicity of the CO $2\pi^*$ orbital by reducing metal retrodonation.

Any complex containing *cis* organyl and carbonyl ligands should therefore be considered to be potentially in tautomeric equilibrium with the corresponding acyl complex. Any acyl complex should be considered to be in equilibrium with the corresponding *cis* organyl/carbonyl complex, *but only if* a situation arises where the metal centre may become coordinatively unsaturated, *e.g.* thermal or photochemical dissociation of a ligand which is *cis* to the acyl ligand.

Applications of the migratory insertion process are numerous and many will be met in subsequent sections. One example, however, which has been particularly successful is the Monsanto process for the carbonylation of methanol to provide ethanoic (acetic) acid (Figure 3.27).

The catalysts for the Monsanto process are hydrogen iodide and the complex $[RhI_2(CO)_2]^-$. Hydrogen iodide converts methanol into iodomethane, which (unlike methanol) is highly prone to oxidative addition reactions with electron-rich metal centres (see also Figure 2.20). The rhodium complex has features typical of many catalysts: two easily accessible and favourable oxidation states with their associated geometries square planar (d^8 ML_4) and octahedral (d^6 ML_6). These are readily interconvertible *via* the processes of oxidative addition and reductive elimination. Oxidative addition of iodomethane provides a *cis* methyl/carbonyl complex (18VE) in equilibrium with a 16VE acyl complex. The subsequent coordination of CO blocks the reverse de-insertion and this is followed by reductive elimination of ethanoyl (acetyl) iodide. This might re-add to the metal were it not rapidly hydrolysed to ethanoic

Controlled, and ideally catalytic, C–C bond formation in its many and various forms is the crowning achievement of modern organometallic chemistry, and its continuing preoccupation.

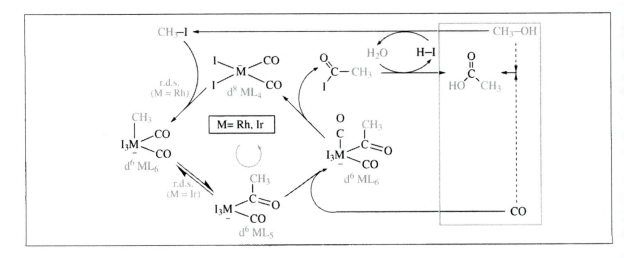

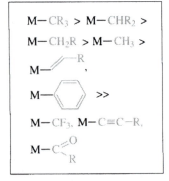

Figure 3.27 Catalytic carbonylation of methanol (r.d.s. = rate-determining step)

Figure 3.28 Empirical aptitude for migratory insertion

acid and hydrogen iodide by the water present (from iodomethane formation).

The process works exceedingly well for the rhodium, iodide and methanol combination, although it has been extended to iridium, other halides and longer chain alcohols with some success. The success with methanol and iodide as the halide lies in the particularly rapid rates of oxidative addition of iodomethane compared with other alkyl halides (weak C–I bond strength, fast S_N2 substitution). This is the rate-limiting step (rate-determining step, r.d.s.) in the case of rhodium. The iridium complex $[IrI_2(CO)_2]^-$ will also mediate the process; however, it is noteworthy that in this case the rate-limiting step is the migratory insertion. Metal–carbon bond strengths generally increase down a triad (Table 1.5); thus for iridium the oxidative addition reaction is more facile (Ir–C bond formation as the transition state is approached). The migratory insertion reaction, however, is less favoured (Ir–C bond disruption on approach to the transition state).

Table 1.5 also holds one key for interpreting the empirical order of migratory aptitudes for various organyl groups when the BDEs for various $RMn(CO)_5$ complexes are considered. Since it is the Mn–R bond which is broken in the migratory insertion reaction, the greater the BDE, the less favourable the migratory insertion reaction. Figure 3.28 shows an empirical ordering of σ-organyl groups.

Generally, any factor that enhances the metal–carbon bond strength may discourage migratory insertion. This includes sp^2 or sp hybridization at carbon, which adds a (variable) π-component to the metal–carbon bond. However, it must be stressed that these are thermodynamic arguments. In terms of kinetic activation, the approach to the three-membered transition state may be assisted by further orbital interactions available to unsaturated σ-organyl ligands. The hydride ligand presents a special case in that it is very rarely *directly* observed to undergo migratory insertion reactions with carbonyl ligands. There is, however, considerable indirect evidence that such processes operate (Figure 3.29). Indeed, such a process is considered crucial to the catalytic (heterogeneous) Fischer–Tropsch synthesis whereby CO is reduced by hydrogen to provide a range of organic molecules of the form $C_xH_yO_z$. The problem in observing the reaction lies in appreciating the interplay of thermodynamic and kinetic factors; in general, the BDEs for M–H bonds exceed those of other σ-organyls (Table 1.5). Thus although hydrido-carbonyl complexes may well be in rapid equilibrium with formyl tautomers, the position of this equilibrium (and hence the amount of any spectroscopically observable species) will lie strongly towards the hydride isomer.

A variety of ligand substitution reactions of hydrido-carbonyl complexes appear to proceed *via* associative activation (rate dependant on the nucleophilicity and concentration of the incoming ligand), pointing towards the intermediacy of coordinatively unsaturated formyl species (*e.g.* Figure 3.29a). Thus, for example, $[FeH(CO)_4]^-$ undergoes ligand substitution reactions far more rapidly than $Fe(CO)_5$, $[Fe\{C(=O)Me\}-(CO)_4]^-$ or $[Fe(CO)_4]^{2-}$. The hydride migratory insertion route appears most important for complexes that have weak metal–hydride bonds [*e.g.* $MnH(CO)_5$, $FeH_2(CO)_4$, $CoH(CO)_4$]. Complexes with stronger

Figure 3.29 Supporting evidence for migratory insertion reactions of hydride ligands

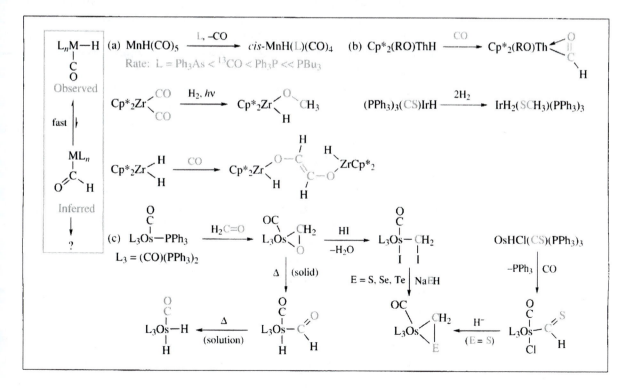

metal–hydride bonds appear less reactive, *e.g.* ReH(CO)$_5$ undergoes substitution *via* radical intermediates, and no deuterium isotope effect is encountered in the substitution reactions of Mo(H/D)(CO)$_3$Cp, consistent with negligible Mo–H bond disruption in the rate-limiting step. One situation where the equilibrium may be shifted towards observation of the formyl tautomer is provided by the carbonylation of the complex ThH(OR)Cp*$_2$ (Figure 3.29b). In this case the highly oxophilic thorium centre 'traps' the oxygen of the formyl complex in a bidentate (3VE) coordination.

3.6.4 Carbonyl Metallates

It was noted above that situations could be contrived in which the oxygen atoms(s) of carbonyl metallates may be attacked by hard and sterically encumbered electrophiles (Figure 3.19). These are illustrative exceptions to the far more general observation that the metal centres of carbonyl metallates typically act as potent nucleophiles. This reverses the classical view of metals centres as Lewis acids. Carbonyl ligands can make metals into strong bases and nucleophiles by stabilizing unusually low oxidation states and the associated high electron density. It is this feature that underlies their enormous synthetic utility. Because they are highly reduced and ligated by π-acids, the EAN rule is strictly obeyed almost exclusively for metal carbonylates. The nucleophilicity of carbonyl metallates is illustrated for three typical examples: Na[Mn(CO)$_5$] and [Na(HMPA)$_x$]$_3$[Mn(CO)$_4$] (Figure 3.6) and Collman's reagent Na$_2$[Fe(CO)$_4$]·1.5dioxane (Figure 3.30). The manganese examples illustrate the formation of bonds between manganese and a variety of main group elements (Groups 13–17) and transition metals. Figure 3.30 illustrates the use of carbonyl metallates in stoichiometric organic synthesis. The acyl ferrate originates from (i) the reaction of Collman's reagent with acyl halides; (ii) attack upon Fe(CO)$_5$ by a suitable carbanionic nucleophile (Figure 3.20); or (iii) from the migratory insertion of the simple alkyl derivative (prepared from the corresponding alkyl halide) upon carbonylation. Electrophilic cleavage of either the alkyl or acyl species provides access to aldehydes, ketones, carboxylic acids, acid halides, esters and amides. Certainly these are all functional groups for which copious alternative synthetic procedures exist in classical organic chemistry. The strength of the Collman protocol lies in the fact that much classical organic carbonyl chemistry involves the groups R and R* being introduced in nucleophilic form; the present approach allows these groups to be introduced in nucleophilic (Figure 3.20) *or* electrophilic form. The use of very hard alkylating agents and polar solvents (*e.g.* HMPA) may result in alkylation at oxygen (rather than iron) to provide alkoxycarbenes which themselves have a rich organic chemistry (Chapter 5).

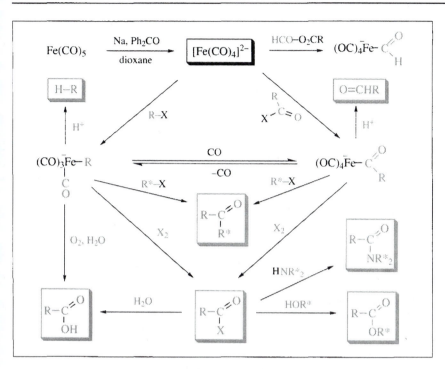

Figure 3.30 Applications of Collman's reagent (X = halogen)

3.7 Ligands Related to Carbon Monoxide

There exists a series of ligands which are isoelectronic with CO (Figure 1.14) and which might be expected to show some parallels in bonding and reactivity. Many such parallels do indeed arise; however, it is the divergences which intrigue. Whilst the frontier orbitals of these various ligands might be similar in shape, number and occupancy, the various electronegativities of the elements concerned lead to significant differences in reactivity. Thus, for example, whilst electrophilic attack at the oxygen atom of coordinated CO is rare, cyanides (CN), isocyanides (CNR) and chalcocarbonyls (CA: A = S, Se, Te) appear more prone to attack at the β-atom (Figure 3.19). Chalcocarbonyls, in particular CS, have a rich coordination chemistry, despite the limitation that CS itself is not isolable. Ingenious methods have been developed for the construction of this ligand within the protective environment offered by coordination to a metal (Figure 3.31). Many more examples will be met in later sections. The chalcocarbonyl ligands are stronger π-acids than CO owing to the π* orbitals of the CA decreasing in energy on descending Group 16. This is reflected in a greater propensity for CS to enter into migratory insertion processes (Figure 3.29) and reactions with nucleophiles.

The stabilization of reactive molecules via coordination is a recurrent theme of organotransition metal chemistry.

Isocyanides, in contrast, are much poorer π-acids and stronger σ-donors than CO. Accordingly, they appear less able to stabilize

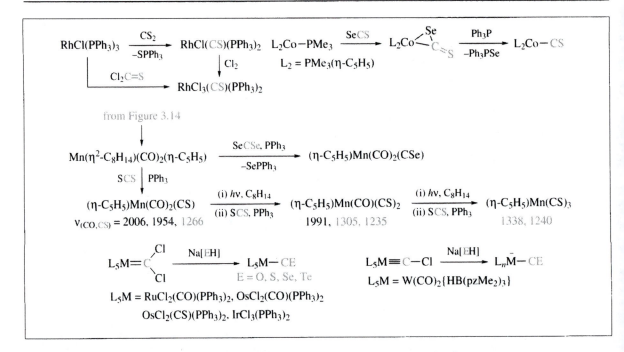

Figure 3.31 Selected syntheses of chalcocarbonyl complexes

In addition to their toxicity, their malodorous nature and tendency to polymerize have retarded the study of isocyanides.

negative or low oxidation states. The field of isocyanide metallates is, however, beginning to grow slowly. Analogues of some of the more familiar carbonyl metallates have now been prepared, *e.g.* [Co(CNR)$_4$] and [Ru(CNR)$_4$]$^{2-}$, as well as zerovalent binary isocyanide complexes, *e.g.* Fe(CNR)$_5$, Fe$_2$(CNR)$_9$ and Co$_2$(CNR)$_8$. The comparative influence on metal basicity of CO and isocyanide ligands is indicated by the observation that Fe(CO)$_5$ is only protonated by concentrated H$_2$SO$_4$ (Figure 2.13), whilst Fe(CNBu^t)$_5$ will deprotonate WH(CO)$_3$Cp (pK_a = 16.1).

4

σ-Organyls

Aims

This chapter introduces complexes with metal–carbon σ-bonds, *i.e.* σ-organyls. By the end of this chapter you should have an appreciation of:

- Synthetic routes applicable to a particular complex, and the associated reactivity options
- The various decomposition routes of transition metal σ-organyls
- How such processes may be harnessed for the controlled functionalization of metal σ-organyls
- Representative situations where these principles may be built into productive catalytic cycles (see also Chapter 6)

4.1 Introduction

Curiously, the simple metal–carbon single bond was amongst the last to be embraced by organometallic chemists. This situation arose from premature ideas about the presumed instability of transition metal alkyls, based on the very few examples known by the middle of the 20th century. There is nothing inherently unstable about metal–carbon σ-bonds from a thermodynamic perspective; some transition metal alkyls and aryls have BDEs in excess of those for main group elements (Table 1.5). The key lies in appreciating the kinetic instability (lability) of many transition metal–carbon σ-bonds, *i.e.* the many ways in which they may decompose, through low-energy routes less accessible to main group (closed d-shell) elements. Once understood, practical use can be creatively made of such decomposition routes, in imaginative applications to organic synthesis. Alternatively, situations may be engineered to allow

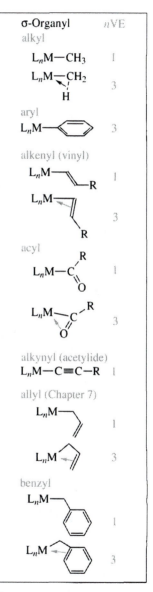

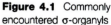

Figure 4.1 Commonly encountered σ-organyls

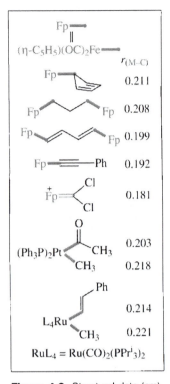

Figure 4.2 Structural data (nm) for σ-organyls

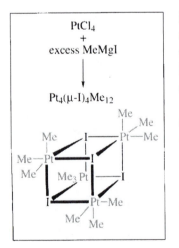

Figure 4.3 Pope's complex

Less sterically cumbersome alkyls can bridge two transition metal centres, *cf.* alkyls of Groups 1, 2 and 13 (Figures 1.1b, 4.1).

the isolation of kinetically (or thermodynamically) stabilized σ-organyl complexes. The various classes of σ-organyls to be discussed are depicted in Figure 4.1. We will be concerned primarily with mononuclear examples; however, many of these organyls can bridge two or more metal centres. Also noteworthy are the ways in which some organyl ligands can provide more than 1VE, if required by the coordinative unsaturation of a metal centre (*e.g.* alkenyls, acyls and allyls).

Of particular interest are σ-organyls having empty orbitals of π symmetry (with respect to the M–C vector) available for overlap with occupied metal orbitals of π symmetry (t_{2g}). Such ligands, *e.g.* alkenyls (vinyls), acyls and alkynyls (acetylides), can in principle supplement their σ-bonding with a π-retrodative component. This may be reflected in an increased M–C bond strength (Table 1.5), a decrease in the M–C bond length (Figure 4.2) or in restricted rotation about the M–C bond, as sometimes observed in VT NMR experiments.

More importantly, much chemical reactivity points towards these interactions being important, *e.g.* alkenyl and acyl complexes tend to display enhanced nucleophilicity at the C_β or O atoms, respectively, suggesting transfer of electron density from the metal (see later).

4.2   Selected Classes of σ-Organyls

4.2.1 Alkyls

Pope's landmark complex $\{PtMe_3I\}_4$ (1907, Figure 4.3) predated a four-decade hiatus in the chemistry of metal alkyls, but with hindsight holds the key to the stability of metal alkyls. Each platinum centre has an octahedral d^6 coordinatively saturated geometry. By far the majority of transition-metal σ-organyl decomposition routes require the availability of a vacant coordination site adjacent (*cis*) to the organyl and an electron count of <18VE. In the interim, however, homoleptic alkyls (neutral and ionic) have been prepared (Figure 4.4).

These are generally limited to what are termed kinetically stabilized alkyls, *i.e.* those devoid of protons β to the metal (Figure 4.5). These also include norbornyl and adamantyl examples since decomposition *via* β-metal hydride elimination (see below) would require the installation of an alkenic bond between the α and β carbons of the precursor alkyl. This is precluded for these two alkyls because of the prohibitive strain associated with forming a double bond to a bridgehead atom within a small cage structure (Brendt's rules). The tetrakis(norbornyl) complexes are also remarkable because some uncharacteristic oxidations states can be attained, *e.g.* Cr(IV), Co(IV) (low-spin $e^4t_2^0$). The second factor which may confer stability is steric bulk; bimolecular decomposition routes are thereby discouraged.

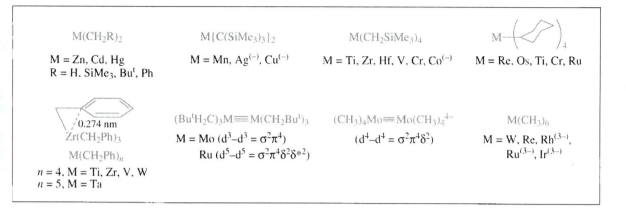

There are various ways in which the coordinative unsaturation of homoleptic alkyls may be accommodated. Coordinatively unsaturated alkyls may show M–C_α–C_β angles greater than the value expected (C sp^3: 109°), and when the protons can be located crystallographically, one shows an agostic interaction with the metal centre (Figure 2.24). Similarly, protons β to the metal may agostically stabilize the alkyl; however, this is less commonly encountered in favour of complete β-M–H elimination. In solution, such agostic interactions *may* be inferred from a shift to higher field of the ^{1}H NMR resonance for the CH_α (or CH_β) proton and/or a decrease in the value of $^1J(^{13}C-^1H_\alpha)$.

In the case of benzyl complexes, an interaction with the *ipso* carbon may occur, or even with both the *ispo* and one *ortho* carbon, effectively an η^3-allyl coordination (Figure 4.1, Table 1.1). Some stable alkyls may contain chelating donor groups which confer stability through the chelate effect.

Mixed ligand alkyl complexes, with kinetically or thermodynamically stabilizing co-ligands, are far more numerous and examples have already been met. The more general synthetic routes to alkyl complexes are summarized in Figure 4.6, many of which apply to σ-organyls in general. Amongst these, the transmetallation (transfer of the σ-organyl between metals) from often commercially available organometallic reagents of Groups 1, 2, 12, 13 and the heavier elements of 14 are the most commonly employed. Such reagents effectively deliver the organyl ligand in a nucleophilic (carbanionic) form in substitution reactions with complex halides, alkoxides or amides (nucleophilic substitution or metathesis). These equilibria are usually dictated by the thermodynamics of the alkyl transfer; transfer from a more electropositive metal to a more electronegative one generally results. However, the heat of formation of the resulting halide must be considered; for ionic halide formation (*e.g.* Groups 1 and 2) this may provide a substantial component of the driving force. The more 'carbanionic' and highly nucleophilic

Figure 4.4 Selected homoleptic alkyls

Caution should be exercised in interpreting solution NMR data since dynamic processes may operate.

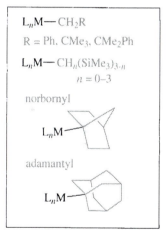

Figure 4.5 Kinetically stabilized alkyls

Particular combinations of **geometry** and **electronic configuration** may also confer stability on some alkyls, e.g. d^0 octahedral [Cr(CHCl$_2$)(OH$_2$)$_5$]$^{2+}$ and d^8 square planar Ir(Me)(CO)(PPh$_3$)$_2$.

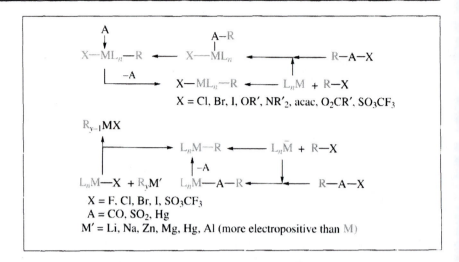

$$X = Cl, Br, I, OR', NR'_2, acac, O_2CR', SO_3CF_3$$

$$X = F, Cl, Br, I, SO_3CF_3$$
$$A = CO, SO_2, Hg$$
$$M' = Li, Na, Zn, Mg, Hg, Al \text{ (more electropositive than } M)$$

Figure 4.6 σ-Organyl transfer

organometallics of the very electropositive elements of Groups 1 and 2 occasionally lead to complex multiple metatheses, or single-electron transfer (SET) reactions (reduction). In such cases, control may be achieved by use of the milder reagents of Groups 12 and 13. In the case of coordinatively saturated carbonyl halide complexes, nucleophilic attack may occur first at a carbonyl ligand followed by loss of halide and migration of the organyl to the metal centre (see Chapter 3).

For electron-rich metal substrates, oxidative addition of alkyl halides offers a means of delivering the alkyl in electrophilic form, although it ultimately adopts a $^{\delta+}M-C^{\delta-}$ bond polarity as a result of the relative electronegativities of carbon and the metal centre. For this purpose, the utility of the halide increases for $X = F < Cl < Br < I$ (decreasing C–X bond strength). Although these reactions are most commonly simple S_N2 in nature, radical routes may also be travelled. The routes shown in Figure 4.7 involve modification of pre-coordinated ligands and these are revisted in more detail in sections dealing with the precursor ligand.

4.2.2 Aryls

Aryls are amongst the oldest examples of σ-organyls to be investigated. Although there may be a modest π-component contributing to the thermodynamic bond strength, these π-interactions may also facilitate approaches to the various transition states for decomposition. Over 80 years ago, Hein investigated the synthesis of homoleptic aryls of trivalent chromium. The ultimate products (Figure 4.8) contained *hexahapto* arene coordination, although at the time this was not recognized as such (Chapter 7). The related complex $CrPh_3(THF)_3$ can be isolated; however, the corresponding complex formed in diethyl ether (a more common solvent in Hein's day), $CrPh_3(OEt_2)_3$, is far less stable. Figure 4.8 also

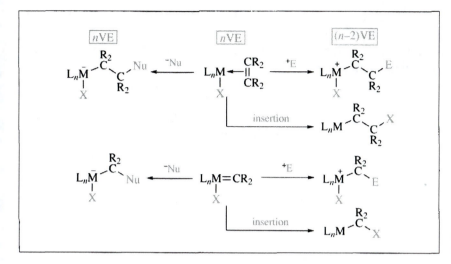

Figure 4.7 σ-Alkyl ligands *via* modification of other ligands

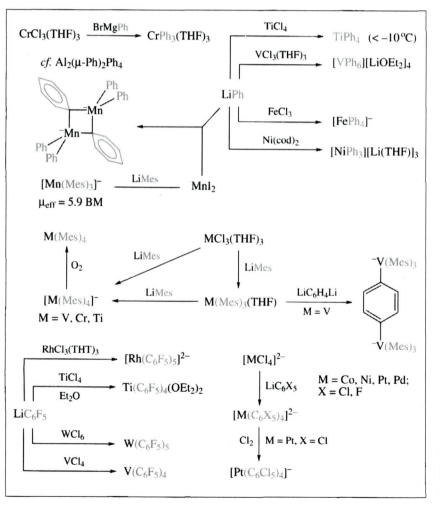

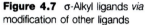

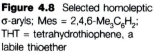

Figure 4.8 Selected homoleptic σ-aryls; Mes = 2,4,6-Me₃C₆H₂; THT = tetrahydrothiophene, a labile thioether

depicts further isolable examples of homoleptic aryls, amongst which the C_6X_5 (X = F, Cl) group appears to provide particularly robust derivatives, through a combination of steric (kinetic) and electronic (thermodynamic) stabilization.

Reductive elimination of biaryls is a common decomposition route (Figure 4.9a), as illustrated by the osmium example (Figure 4.9b) in which the biaryl remains *hexahapto* coordinated. Alternatively, benzyne complexes may result from β-C–H or β-C–Br abstraction processes (see below).

Benzyne ligands might also be described as phenylene-1,2-diyls $(C_6H_4^{2-})$.

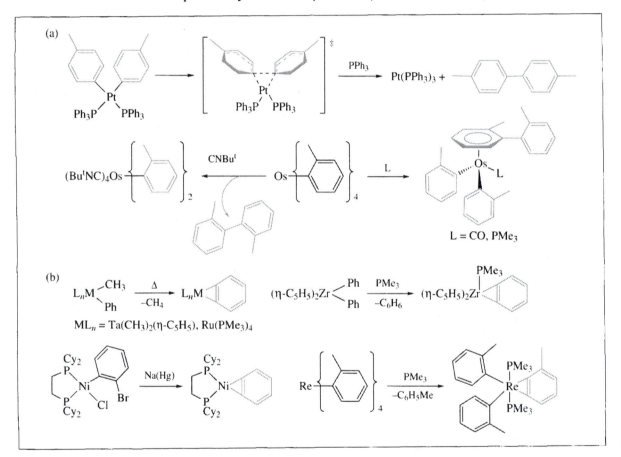

Figure 4.9 Decomposition of aryl complexes *via* (a) biaryl or (b) aryne formation

The facility of arene reductive elimination underpins numerous C–C, C–O and C–N bond-forming reactions, which may be catalysed by late transition metals, in particular palladium (Figure 4.10). Although there are many variants, the general reaction scheme involves introduction of the aryl in electrophilic form *via* oxidative addition of an aryl halide (or sulfonate), substitution of the palladium halide by a nucleophile (which may also be carbon based) followed by reductive elimination. It is noteworthy that nucleophilic aromatic substitution in the absence of such catalysts can be difficult.

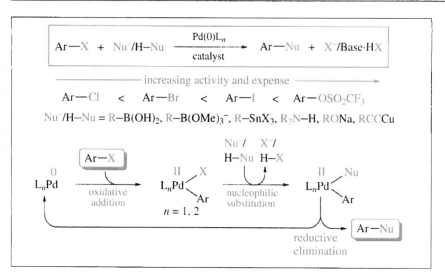

Figure 4.10 Generalized catalytic scheme for palladium-catalysed aryl halide (Ar-X) functionalization

4.2.3 Alkenyls (Vinyls)

As with aryls, the sp² hybridization of σ-alkenyls may contribute to both an increased BDE and also access to decomposition routes; homoleptic examples, *e.g.* Zr(CPh=CMe₂)₄, are particularly rare. Decomposition of *cis*-bis(alkenyl) complexes *via* reductive elimination of buta-1,3-dienes is occasionally observed; however, the butadiene *may* remain coordinated (Figure 4.11).

Other than simple transmetallation approaches, the majority of synthetic routes to σ-alkenyls involve the functionalization of alkynes, *via* reactions of the coordinated alkyne with suitable nucleophiles or electrophiles or alternatively by the insertion of the alkyne into an adjacent M–H or M–C bond. These reactions (Figure 4.12) will be revisited in Chapter 6.

One significant feature of alkenyl ligands is that in addition to simple

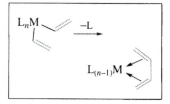

Figure 4.11 Butadiene reductive 'elimination'

Figure 4.12 Synthesis of σ-alkenyl ligands *via* alkyne modification

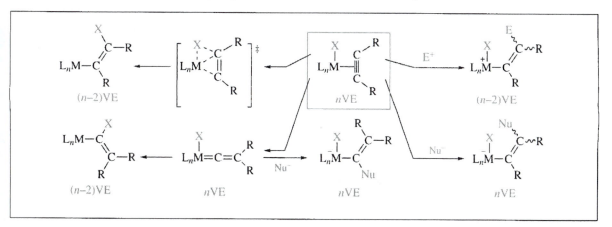

σ-coordination (1VE donation), they offer the possibility of σ–π or η² coordination (3VE; Figure 1.6) to metal centres that would otherwise be coordinatively unsaturated. In some instances, such ligands may rearrange to alkylidyne ligands (also 3VE, Figure 4.13; see also Chapter 5). The ability to contribute 3VE also allows alkenyl ligand to bridge two metal centres (Figure 4.14).

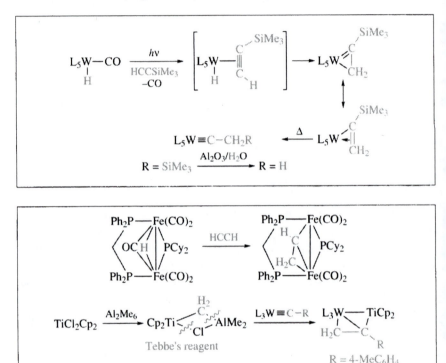

Figure 4.13 σ–π Vinyl–alkylidyne interconversion (see Figure 2.11); L₅W = W(CO){HB(pzMe₂)₃}

Figure 4.14 Bridging alkenyl (vinyl) ligands

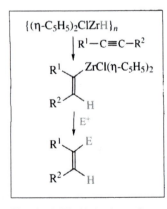

Figure 4.15 Hydrozirconation with Schwartz's reagent (see Figure 2.29b)

Schwartz's reagent, {ZrHClCp₂}$_n$, (Figure 2.29b), readily (and often regioselectively) hydrozirconates alkynes to provide σ-vinyl complexes from which the vinyl ligand may be cleaved by a wide range of electrophiles, in a manner reminiscent of hydroboration chemistry (Figure 4.15).

4.2.4 Acyls

These ligand have already been met (Chapter 3), including the three key synthetic routes: migratory insertion of σ-organyl/carbonyl complexes, electrophilic attack at electron rich metals by acyl halides or anhydrides and nucleophilic attack at coordinated CO (Figure 4.12).

The reverse of migratory insertion, α-organyl elimination, is also fundamental to the reactivity of acyls. Further synthetic approaches are summarized in Figure 4.16 (see also Figure 3.30).

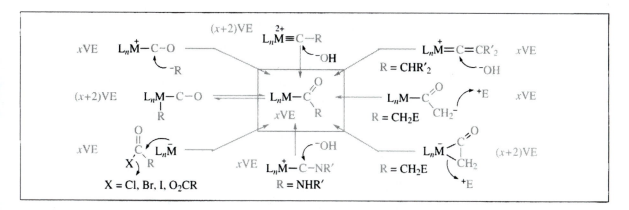

A further feature of acyl ligands is the enhanced acidity of protons on the carbon adjacent to the carbonyl (*cf.* enolate chemistry), a property shared with alkyl-substituted Fischer-type carbenes (Chapter 5). Figure 4.17 illustrates this point, including an application employing the chiral auxiliary 'Fe(CO)(PPh₃)Cp'.

Figure 4.16 General routes to acyl ligands

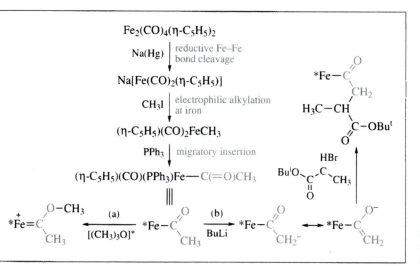

Figure 4.17 Acyl functionalization by (a) *O*-alkylation and (b) β-C–H deprotonation

Heteroacyls, including alkoxycarbonyls [L_nM–C(=O)OR] and carbamoyls [L_nM–C(=O)NR₂], fall within this category with the exception that the carbon–heteroatom bond may occasionally be cleaved by Lewis (or Brønsted) acids (α-abstraction) to generate a carbonyl ligand, a reaction which is atypical of hydrocarbon-substituted acyl ligands. The oxygen atom of acyls, in particular anionic acylates, may display nucleophilic character, especially towards hard electrophiles, a feature that led to the first examples of transition metal carbenes by alkylation (Figure 4.17a; see also Figure 5.4).

4.2.5 Alkynyls (acetylides)

For a long time, σ-alkynyls, by analogy with isoelectronic cyanides (Figure 1.14), were viewed from a classical coordination chemist's perspective, rather than as synthetically versatile organometallic functional groups. This by no means remains the case owing primarily to two principles. Firstly, many catalytic approaches to alkynes now proceed *via* alkynyl complexes of palladium and/or copper. Secondly, the facile conversion of alkynyls to vinylidenes (Chapter 5) presents a versatile and reactive M=C=C system which provides access to alkenyl, alkylidene, acyl, alkylidyne, carbonyl and alkyl species *via* the complex reaction manifold shown in Figure 4.18. This is an attempt to generalize a vast amount of chemistry in a unified scheme; the *ultimate* products of these reactions tend to be very metal-specific but clearly have enormous synthetic potential. The crucial alkyne–vinylidene interconversion will be discussed in detail in Chapter 5.

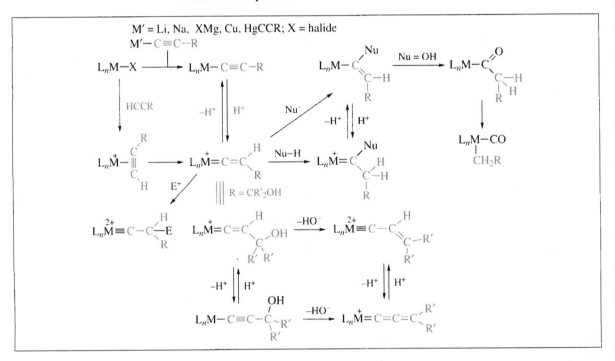

Figure 4.18 Organometallic ligands arising from the modification of σ-alkynyls

Alkynyl ligands are only modest π-acids; however, this can allow electronic communication between metal centres bound to bis(alkynyl)diyl bridging ligands, *e.g.* the diruthenium butadiyne-1,4-diyl complex shown (Figure 4.19) undergoes *four* single 1VE oxidations. If the metals were not in electronic communication, oxidation of one iron centre would not significantly affect the redox potential of the second, *i.e.* both would oxidize at the same potential.

The ability of alkynyl groups to bridge two metals may be employed in the construction of heterometallic complexes in a controlled manner (Figure 4.20; Chapter 6). Many metals will react directly with terminal alkynes or acetylide salts under basic conditions; however, use is often made of copper alkynyls as *trans*-alkynylating reagents [usually generated *in situ* with catalytic amounts of Cu(I) salts].

4.3   Reactions of σ-Organyls

As noted above, many transition metal σ-organyl complexes are kinetically unstable owing to low-energy decomposition routes, which become available if the metal centre is coordinatively unsaturated. The various decomposition routes will be discussed in turn within the wider context of their reactivity.

4.3.1   Homolytic Cleavage

The homolytic cleavage (formation of organyl radicals) of transition metal σ-organyls is not generally a favourable decomposition route under mild conditions. However, it does arise in some situations, the most celebrated being in the biochemistry of vitamin B_{12} where a number of reactions are mediated by its ability to generate adenonsyl radicals. More generally, however, the majority of transition metal σ-organyl complexes have metals in low oxidation states, and quite often also have π-acceptor co-ligands. These factors favour the adoption of low-spin diamagnetic complexes, such that loss of an organic radical and formation of a paramagnetic metal radical is less favoured than other lower energy (diamagnetic) processes. The use of cobalt σ-organyl porphyrin complexes

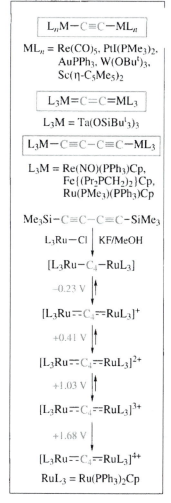

Figure 4.19 Yne-diyl and diyne-diyl ligands

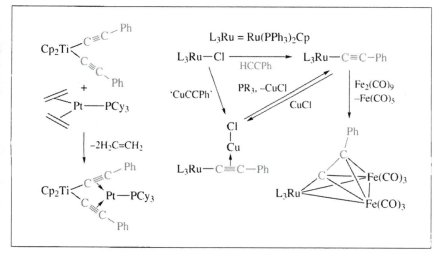

Figure 4.20 Alkynyl-bridged bi- and trimetallic complexes

(models for vitamin B_{12}) as catalysts for the living polymerization of methyl methacrylate similarly relies on the ability of these complexes to generate radicals by facile Co–C homolysis.

4.3.2 Reductive Elimination

Complexes with two adjacent (*cis*) σ-organyl ligands, or (more readily) a σ-organyl and a hydride ligand, may decompose to provide hydrocarbon with concomitant (formal) reduction of the metal centre by two units. Either mononuclear (Figure 4.21a) or bimolecular (Figure 4.21b) decomposition routes may operate, depending on the steric and electronic properties of the particular metal complex. This has been most

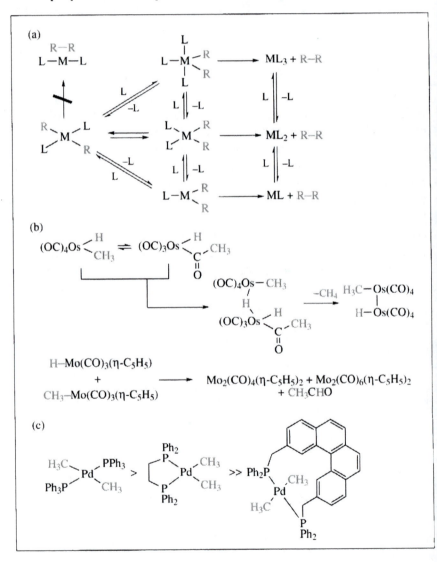

Figure 4.21 Alkane reductive elimination

intensely studied for d^8 square planar diorganyls of Group 10, where the actual mechanism may involve (i) prior dissociation of a co-ligand to generate a more reactive three-coordinate species (dissociative activation), or (ii) addition of a ligand to provide a more sterically encumbered five-coordinate intermediate (associative activation). This is especially true if the initial complex has the two eliminating groups in mutually *trans* positions, and therefore unable to commence elimination from a four-coordinate geometry. This is illustrated by the relative rates of ethane elimination from the three dimethyl complexes shown (Figure 4.21c), the third of which has a bidentate diphosphine which enforces a *trans* geometry.

In general, reductive elimination involving an organyl and a hydride is more facile than that between two organyls; indeed, hydrogenolysis of metal alkyls represents one of the most important cleavage reactions, which is often incorporated within catalytic cycles (Figures 4.22, 6.14). For metals in less than their highest oxidation state, this may proceed *via* oxidative addition of dihydrogen followed by reductive elimination

Figure 4.22 Hydrogenolysis of σ-alkyl ligands

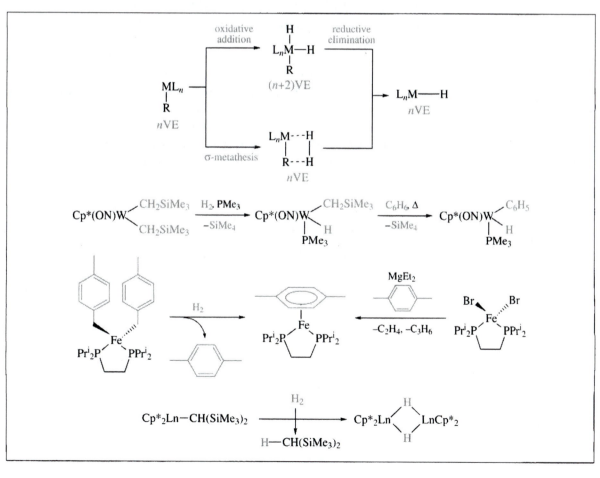

Table 4.1 Bond dissociation enthalpies (kJ mol^{-1})^a

H–H	435
H–Ph	460
H–CH=CH$_2$	452
H–CH$_3$	435
H–CH$_2$Me	410
H–CH$_2$Ph	368

^{a}1 cal = 4.184 J

As in the case of chelated agostic interactions, a shift to high field of the coordinated nuclei is observed in both the ^{1}H and ^{13}C NMR spectra, in addition to a decrease in the typical value of $^1J(^{13}C-^1H)$ (Figure 4.23b).

of a hydrocarbon. For metals in higher oxidation states, the cleavage may alternatively proceed *via* a (highly ordered, $\Delta S^{\ddagger} < 0$) four-membered σ-metathesis transition state. Combining the BDE data given in Table 1.5 with those in Table 4.1 suggests that a thermodynamic impetus for alkyl hydrogenolysis lies in the greater value for M–H *vs.* M–C bonds.

The (microscopic) reverse of organyl-hydride elimination is known as C–H activation and is potentially of enormous technological significance, especially in the case of non-functionalized alkane C–H bonds. The activation of non-functionalized alkanes occurs in nature, *e.g.* the iron-based metalloenzyme methane mono-oxygenase, which selectively converts methane to methanol, albeit *via* non-organometallic routes. Early indications that C–H activation was feasible were provided by coordinatively unsaturated species, which entered into oxidative addition reactions with *aryl* C–H bonds (Figures 2.28c, 4.23a). The C–H activation of simple alkanes, however, was first achieved with the transient complexes 'Ir(L)Cp*' (L = CO, PMe$_3$), generated by photolysis of appropriate precursors in alkane solvents (Figure 2.28d). The alkyl hydride complexes that resulted were somewhat unstable with respect to alkane reductive elimination; however, they could be trapped by conversion to the corresponding alkyl-halo complexes on treatment with a halocarbon (a characteristic reaction of metal hydrides; Figure 2.30). An intriguing and highly promising variant on this process involved the C–H activation of *tert*-butyl alcohol and *tert*-butylamine (Figure 2.28d). Following these seminal results, numerous examples of these processes have now been recognized. Furthermore, alkane complexes have now been spectroscopically observed as intermediates, although these are exceedingly labile species. In the case of unsaturated hydrocarbons (alkenes, arenes),

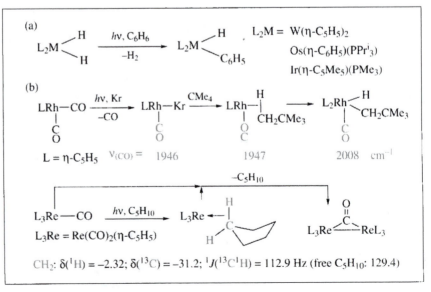

Figure 4.23 C–H activation and alkane coordination

prior coordination of the hydrocarbon through a π-system may occur (see Chapters 5 and 6).

4.3.3 Electrophilic Cleavage

Transition metal alkyls generally retain nucleophilic character at C_α and are thus prone to reactions with electrophiles, usually leading to cleavage from the metal. Mechanistically, two paths may operate, depending on the particular metal centre concerned (Figure 4.24): either direct electrophilic attack at C_α, which leads to inversion of configuration at C_α, or alternatively attack at the metal centre (oxidation) followed by reductive elimination, proceeding with retention of configuration at C_α. These alternatives are depicted in Figure 4.24 in addition to a special case, the insertion of sulfur dioxide. This process is in contrast to conventional migratory insertion (Chapter 3), which involves migration of the organyl to an adjacent ligand and is characterized by retention of configuration at C_α. In the case of SO_2 insertion, however, initial electrophilic cleavage (inversion) results in ion pair formation followed by re-coordination of the sulfinate anion through either sulfur or oxygen. The reverse of this reaction, the de-insertion of SO_2 from sulfinate complexes, also provides a synthetic route to σ-organyls (see also Figure 4.6).

Note that the multiple bond of unsaturated σ-organyls (e.g. alkenyls, alkynyls, acyls, allyls) may present an alternative site to the M–C bond for electrophilic attack.

Figure 4.24 Electrophilic cleavage of alkyl ligands; $FeL_3 = Fe(CO)_2Cp$

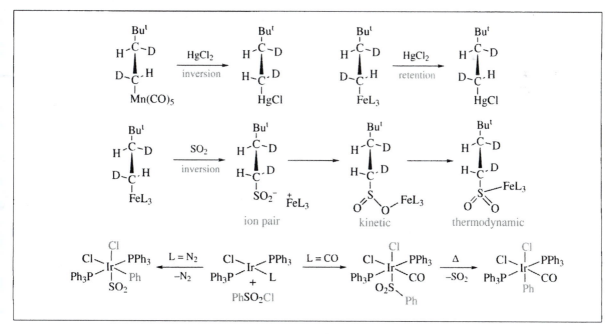

4.3.4 Insertion Reactions

These have already been met in the case of carbonyl–alkyl coupling through migratory insertion (Figures 3.23–3.30). In Chapter 6 the case of alkenes inserting into metal–carbon σ-bonds will be discussed in detail, since it is of crucial importance in catalytic alkene polymerization processes. Figure 4.25 illustrates the general cases of migration to a C_1 and a C_2 ligand and situations where these migrations have been implicated. Parallels with insertion reactions of hydride ligands (Figures 2.30, 2.31) should become apparent, bearing in mind that M–C bonds are typically 60–120 kJ mol^{-1} weaker than M–H bonds (Table 1.5).

Although we are primarily concerned here with migrations to carbon ligands, the general principles should be seen as applying to other unsaturated ligands.

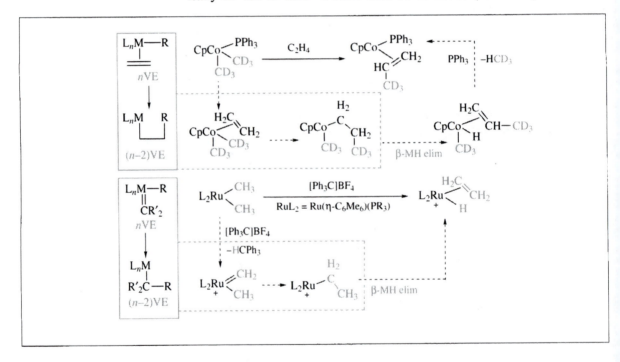

Figure 4.25 Migrations of alkyl ligands to C_1 and C_2 co-ligands

4.3.5 β-Metal Hydrogen (β-M–H) Elimination

This reaction is by far the most common decomposition route for alkyl ligands which possess hydrogen substituents on the carbon β to the metal. The process requires a vacant coordination site (*i.e.* <18VE) *cis* to the alkyl (Figure 2.31). The β-C–H bond may weakly coordinate in an agostic manner (sometimes isolable; Figure 2.24) *en route* to a four-membered transition state involving C–H bond rupture (transfer of the hydrogen to the metal). The resulting alkene may remain coordinated or alternatively dissociate and be substituted by other available ligands. As such, this reaction presents an important synthetic route to hydride complexes (Chapter 2, Figure 2.31). If more than one alkyl ligand is present, the resulting hydride may reductively eliminate alkane with a second alkyl, leading to a reduction of the metal centre. Thus the role of alkylating agents as reductants (*e.g.* Al$_2$Et$_6$) in the synthesis binary carbonyls (Figure 3.11) becomes clear. The geometrical requirement that the vacant site is *cis* to the ethyl ligand is illustrated by the thermal decomposition of *trans*-RuEt$_2$(TPP), which proceeds *via* a radical pathway since the porphyrin (TPP) blocks sites adjacent to the ethyl ligands. Figure 4.26 shows selected examples, including non-organometallic ligands (amides, alkoxides) which are also vulnerable to this decomposition route.

The reactions of metal halides with primary or secondary alcohols represent a common route to metal hydrides, with elimination of aldehydes or ketones, respectively.

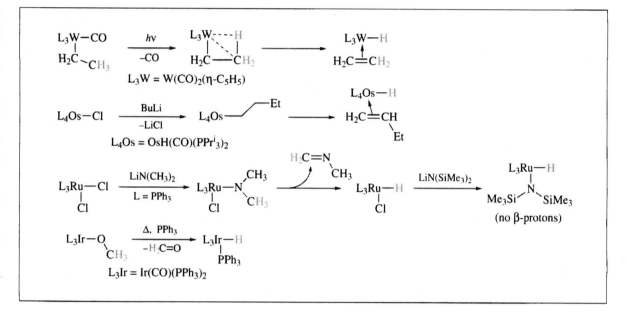

4.3.6 α-Abstraction or α-Elimination

Figure 4.26 Synthetic applications of β-metal–hydride eliminations

The kinetically stabilized alkyls shown in Figure 4.5 do not have the β-M–H elimination route available to them owing to the absence of β-protons. An alternative decomposition route, however, arises when these are bound to highly unsaturated metal centres, involving loss of an α-C–H which may be either transferred (Figure 4.27) to (a) the metal centre itself (α-elimination), (b) a co-ligand or (c) an external basic or electrophilic reagent (α-abstraction).

The complex Ta(CH$_2$Bu^t)$_5$ has yet to be isolated; initial attempts at its preparation led Schrock instead to a remarkable alkylidene (carbene) complex (Figure 4.28). Neopentane was amongst the side products of the reaction and this most likely arises from a concerted α-C–H abstraction by an adjacent neopentyl ligand. In other systems, there is evidence for transfer of the hydrogen to the metal centre and agostic alkyls (Figure 2.24) can be seen as a step *en route* to the α-M–H elimination transition

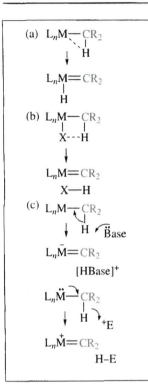

Figure 4.27 (a) α-Hydrogen elimination, (b) intra- and (c) intermolecular hydrogen abstraction

Figure 4.28 Alkyl decomposition *via* intramolecular α-M–H elimination/abstraction

state. These reactions are particularly prevalent for the 4d and 5d elements of Groups 5–7 with low VE counts. The 3d elements, in contrast, are more prone to homolytic cleavage. The α-M–H elimination process requires a somewhat strained three-membered transition state, whilst the β-M–H elimination involves a more relaxed four-membered transition state. In general, β-M–H elimination tends to prevail over α-M–H elimination, although there are (rare) examples where the two processes may compete, *e.g.* the ethene-neopentyl tantalum complex shown in Figure 4.28.

When the α-hydrogen is transferred to an external reagent, it may do so in protic (Figure 4.29) or hydridic form (Figure 4.30a), depending on the nature of the metal centre and the external reagent. Thus the first methylene (CH_2) complex arose from the deprotonation of a cationic methyl complex, which was in turn obtained *via* electrophilic cleavage (Ph_3C^+) of one methyl ligand from $TaMe_3Cp_2$ (Figure 4.29). Microscopic reversibility suggests that since the methylene ligand was produced by removal of an electrofuge (H^+), it will be prone to both the reverse

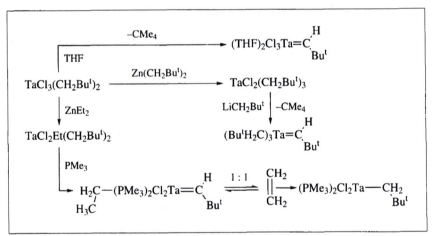

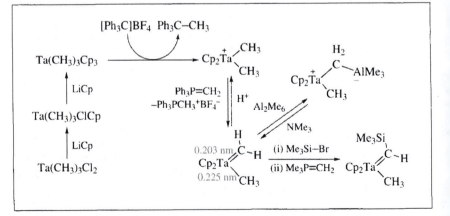

Figure 4.29 Intermolecular α-H abstraction (electrofugic)

reaction (protonation) and reaction with other electrophiles (Me₃SiBr, Al₂Me₆).

In contrast, the late transition metal alkyls Re(CH₂R)(NO)(PPh₃)Cp (R = H, Ph) react with trityl (Ph₃C⁺) salts *via* intermolecular hydride abstraction to provide alkylidene complexes which by microscopic reversibility are prone to reactions with nucleophiles (Figure 4.30a). The reactivity of alkylidene (carbene) complexes will be discussed in the next chapter. Intermolecular α-abstraction is also widely applied to α-alkoxyalkyl and α-haloalkyl ligands, with a range of Lewis acids being suitable (Figure 4.30b).

The generation of cationic alkyl complexes by **electrophilic de-alkylation** of peralkyls using R₃C⁺ or R₃B reagents is *now* a key entry route into molecular **Ziegler–Natta**-type catalysts (see Chapter 6).

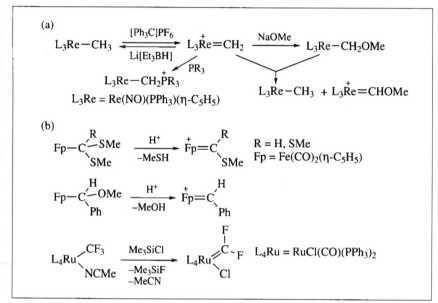

Figure 4.30 Intermolecular α-M–H abstraction (nucleofugic) (see also Figure 5.17)

4.3.7 Remote C–H Functionalization

In addition to α- and β-C–H activation, the possibility occasionally arises for γ- or even δ-functionalization. This is particularly common for aryl phosphine and phosphite ligands that may undergo metallation of the *ortho*-C–H bond of an aryl substituent. This process may be reversible; however, if a suitable co-ligand is present which can undergo subsequent reductive elimination of the hydride, stable metallacyclic organyls are obtained (Figure 4.31). The formation of metallacyclic alkyls may confer some stability, as does the possibility of increased hapticity, *e.g.* in the case of xylyene ligands (see also Chapter 6).

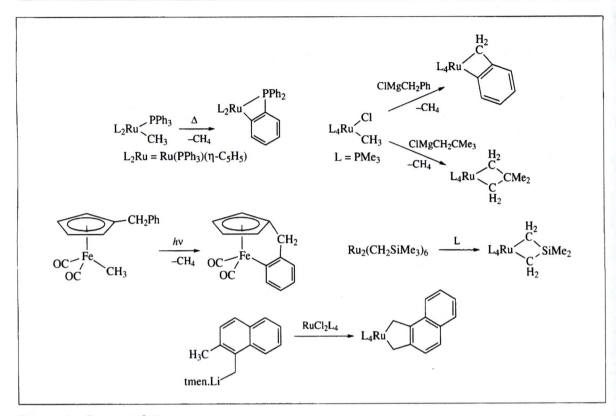

Figure 4.31 Remote (ω) C–H
activation/metallation of organyls

5
Metal–Carbon Multiple Bonding

Aims

By the end of this chapter you should be familiar with the various forms that metal–carbon multiple bonding may take (Figure 5.1), *viz.*

- Alkylidenes (carbenes), $L_nM=CR_2$
- Alkylidynes (carbynes), $L_nM\equiv CR$
- Vinylidenes, $L_nM=C=CR_2$
- Cumulenylidenes, $L_nM=C=C=CR_2$, $L_nM=C=C=C=C=CR_2$

This should include an appreciation of suitable synthetic strategies, typical reactions and selected applications to organic synthesis (stoichiometric and catalytic).

5.1 Introduction

Metal–carbon multiple bonding takes many forms; indeed, carbonyl ligands (Chapter 3) are arguably the simplest. 'Formal' metal–carbon multiple bonds (Figure 5.1) will be considered here; however, some σ-organyl ligands may also feature a degree of metal–carbon multiple bonding (Figure 4.2). The synthesis of such ligands will be discussed, followed by a survey of the various typical reactions that the metal–carbon bond may undergo.

The oxygen atom of acyl metallates may present nucleophilic character (Figures 3.25, 4.16). In 1964, E. O. Fischer showed that protonation of the benzoyl tungstate complex obtained from $W(CO)_6$ and phenyllithium (Figure 3.20) provided the first characterized example of a metal–carbon double bond, *viz.* the hydroxycarbene complex $(CO)_5W=C(OH)Ph$. This labile species decomposed to provide benzaldehyde; however, it could be 'esterified' with diazomethane to provide

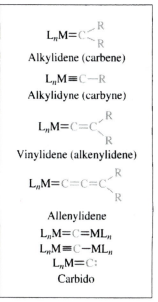

$$L_nM=C\underset{R}{\overset{R}{<}}$$
Alkylidene (carbene)

$$L_nM\equiv C-R$$
Alkylidyne (carbyne)

$$L_nM=C=C\underset{R}{\overset{R}{<}}$$
Vinylidene (alkenylidene)

$$L_nM=C=C=C\underset{R}{\overset{R}{<}}$$
Allenylidene

$$L_nM=C=ML_n$$
$$L_nM\equiv C-ML_n$$
$$L_nM=C:$$
Carbido

Figure 5.1 M–C multiple bonding

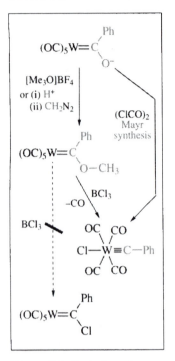

Figure 5.2 Fischer's carbene and carbyne syntheses

the more stable methoxycarbene derivative (Figure 5.2). Almost a decade later, Fischer showed that the reaction of this complex with boron halides did not proceed as expected to provide a halocarbene complex (by analogy with vinyl ethers), but rather the first example of a metal–carbon triple bond, the carbyne complex *trans*-Cl(CO)$_4$W≡CPh.

At the time, such carbene and carbyne complexes must have appeared as exotic curiosities. In the interim, they have become firmly entrenched in the modern synthetic chemist's armoury of synthons. The terms carbene and alkylidene are used interchangeably in the literature; the former is more commonly used for heteroatom functionalized examples whilst the latter typically refers to purely hydrocarbon ligands. In a similar manner, the terms carbyne and alkylidene both enjoy wide useage. Historically, the terms carbene and carbyne call to mind an analogy with the free species CR$_2$ and CR; however, in very few cases does the chemistry or character of these ligands overlap with that of the free (transient) species. This may be understood when one bears in mind two key points. Firstly, the BDE for the metal–carbon double bond of W{=C(OMe)Ph}(CO)$_5$ (*ca.* 359 kJ mol^{-1}; Table 1.5) would appear to preclude simple dissociation [*cf.* 178 kJ mol^{-1} for W–CO in W(CO)$_6$, Table 3.3]. Secondly, the half-lives of simple hydrocarbon carbenes are far shorter than those typical of ligand substitution reactions at transition metal centres. Thus, with a few notable exceptions (see later), free carbenes are neither isolable nor suitable precursors to carbene complexes. One must therefore, in the steps of Fischer, resort to ingenuity to prepare the carbene (or carbyne) ligand within the protective environment of a transition metal coordination sphere.

5.2   Carbenes (Alkylidenes)

A bonding scheme has already been provided for methylene ('CH$_2$') coordinated to a transition metal centre (Figure 1.15). The energies of the frontier orbitals of free CO are fixed; however, the reactivity of coordinated CO may be tuned by variations in the energy of the metal–ligand orbitals. Thereby, the CO may be coordinatively activated towards either nucleophilic or (less readily) electrophilic attack, in addition to ligand coupling reactions. This principle applies equally to carbene (and carbyne) complexes, with the added dimension that now it is also possible to moderate the energies of the carbene frontier orbitals [σ (sp^2) and p$_z$] by varying the carbene substituents. This enormous variability accounts for the wide spectrum of reactivity observed for carbene complexes.

The generalizations in Box 5.1 represent two extreme situations and are referred to in the older literature as Fischer-type and Schrock-type carbenes, in deference to the research groups from which the apparently distinct classes of carbene complex originally arose. This dichotomy

Box 5.1 Carbene Complexes

The following tentative generalizations have some utility:

- For coordinatively saturated carbene complexes ligated by strongly π-acidic co-ligands, with the metal in a comparatively low formal oxidation state, the carbene carbon typically shows electrophilic character. Accordingly, the less reactive examples of these typically bear one or two π-dative carbene substituents (OR, NR_2, SR).
- For coordinatively unsaturated carbene complexes ligated by strong (σ, π) donor co-ligands, with the metal in a higher formal oxidation state, the carbene carbon typically shows nucleophilic character. By far the majority of these involve substituents with no π-dative capacity (H, CR_3, SiR_3).

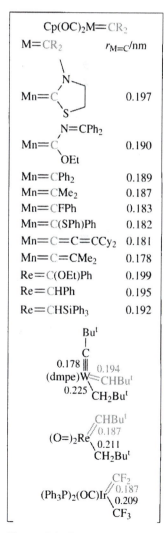

reflected the historical development of two extreme ends of a spectrum, which has in the interim been interspersed with sufficient counter or intermediate examples to justify a more unified view. With regard to carbene substituents, heteroatom groups capable of π-donation into the empty carbene p_z orbital compromise the metal–carbon bond, leading to lengthening (Figures 1.16, 5.3). Halocarbenes present a special case in that the substituents are electronegative and very poor π-donors; unless bound to particularly electron-rich metal centres, halocarbenes are especially electrophilic.

5.2.1 Characterization of Carbene Complexes

A formal metal–carbon double bond should in principle be reflected in the metal–carbon bond length. This is generally the case, although it must be noted that metal–carbene bond lengths may span a wide range due in part to the effects indicated in Figure 1.16. Figures 1.16, 4.2, 4.29 and 5.3 show structural data for selected carbene complexes, amongst which, those for the complex $W(CH_2R)(=CHR)(\equiv CR)(dmpe)$ (R = Bu^t) provide the most convincing (and elegant) illustration. NMR spectroscopy is very useful for characterizing carbene complexes in that the ^{13}C resonance occurs to low field (200–350 ppm). The resonance is moved to higher field by heteroatom carbene substitutents. For secondary carbene complexes ($L_nM=CHR$), ^1H NMR spectroscopy typically reveals a resonance between 10 and 22 ppm, and $^1J(^{13}C-^1H)$ couplings evident in ^{13}C NMR spectra are consistent with sp^2 hybridization at carbon. It should be noted that some coordinatively unsaturated alkylidenes may

Figure 5.3 Structural data for alkylidene complexes

enter into agostic interactions with the metal, leading to an opening of the M=C–R angle, accompanied by a decrease in the $^1J(^{13}C-^1H)$ values (70–100 Hz) and the frequency of the ν_{CH} IR absorption ($\nu_{M=C}$ is seldom unambiguously identified). This agostic coordination is particularly common for coordinatively unsaturated alkylidene complexes of the earlier transition metals. Finally, variable temperature NMR (VT NMR) measurements may reveal barriers to alkylidene rotation (Table 1.3).

5.2.2 Synthesis of Carbene Complexes

Fischer's acylate O-alkylation remains a widely employed method of carbene synthesis, the generality of which is indicated in Figure 5.4. There are, however, many alternative approaches. The case of α-elimination or α-abstraction (Figures 4.27, 4.28, 5.5) is particularly important for unsaturated alkyls of Group 4–7 metals, but less relevant for later transition metals which typically have higher d occupancies.

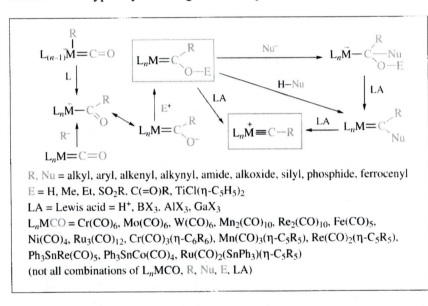

R, Nu = alkyl, aryl, alkenyl, alkynyl, amide, alkoxide, silyl, phosphide, ferrocenyl
E = H, Me, Et, SO_2R, C(=O)R, $TiCl(\eta-C_5H_5)_2$
LA = Lewis acid = H^+, BX_3, AlX_3, GaX_3
L_nMCO = $Cr(CO)_6$, $Mo(CO)_6$, $W(CO)_6$, $Mn_2(CO)_{10}$, $Re_2(CO)_{10}$, $Fe(CO)_5$, $Ni(CO)_4$, $Ru_3(CO)_{12}$, $Cr(CO)_3(\eta-C_6R_6)$, $Mn(CO)_3(\eta-C_5R_5)$, $Re(CO)_2(\eta-C_5R_5)$, $Ph_3SnRe(CO)_5$, $Ph_3SnCo(CO)_4$, $Ru(CO)_2(SnPh_3)(\eta-C_5R_5)$
(not all combinations of L_nMCO, R, Nu, E, LA)

Figure 5.4 Carbonyl-derived carbenes and carbynes

Transition metal complexes of both of these classes of carbene had been prepared previously by modification of other coordinated ligands (Figure 5.7c).

The vast majority of carbene syntheses involve the modification of an existing ligand and some of the more general routes are shown in Figure 5.6, in addition to some less general though intriguing approaches which may subsequently find more generality.

The discovery of stable and isolable free phosphinocarbenes by Bertrand was quickly followed by the isolation of stable imidazolin-2-ylidenes by Arduengo (Figure 5.7). Imidazolylidene complexes of most of the transition metals have since been prepared directly from free carbenes.

The use of stable free carbenes now allows their coordination to be

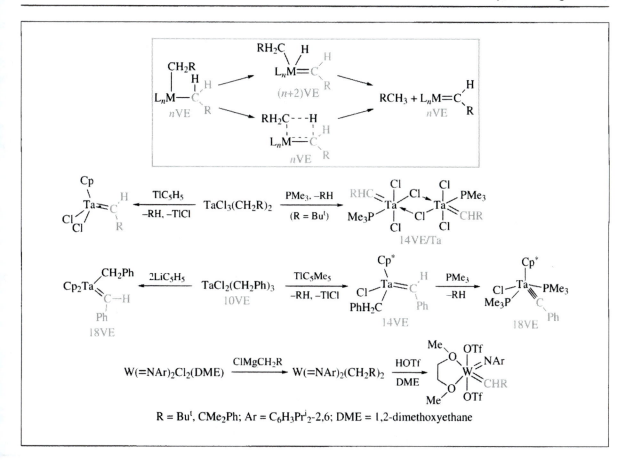

R = Bu^t, CMe$_2$Ph; Ar = C$_6$H$_3$Pr$^i_2$-2,6; DME = 1,2-dimethoxyethane

achieved under milder conditions, thereby greatly increasing their synthetic versatility. These diaminocarbenes have a very modest (if any) retrodative component to the M–C bond and in this respect emulate phosphines both in terms of the net (and tunable) dative capacity and in the potential control of steric factors offered by the variety of possible nitrogen substituents. The reduction in metal–carbon bond order generally results in a loss of M=C based reactivity, to the point that these ligands may be used in place of phosphines as innocent (spectator) coligands in catalytic processes.

As in organic chemistry, diazoalkanes have found wide application as *ultimate* carbene sources. In contrast to the organic chemistry of diazoalkanes, however, the reactions do not proceed *via* free carbenes but rather *via* metal-mediated transformations of coordinated diazoalkanes. In some cases, diazoalkane complexes may be isolated (Figure 5.8). The parent diazoalkane, H$_2$C=N$_2$, has found somewhat less success in the synthesis of terminal methylene complexes L$_n$M=CH$_2$; however, this is primarily due to the lack of any kinetic (steric) or thermodynamic (π-dative) stabilization by carbene substituents. Thus methylene ligands

Figure 5.5 Alkylidene synthesis *via* α-M–H elimination/abstraction (see also Figures 3.28–3.30)

Prior to the isolation of **Arduengo carbenes**, many closely related carbene complexes had also been prepared *via* the thermolytic cleavage of electron-rich alkenes in the presence of various transition metal reagents (Figure 5.6).

Figure 5.6 Selected synthetic routes to carbene complexes

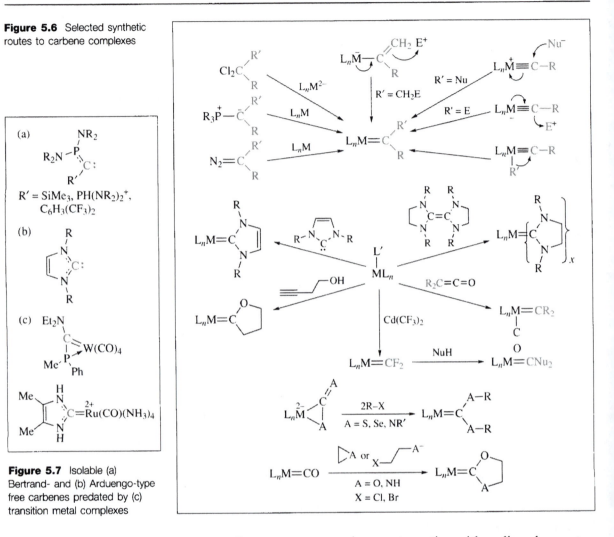

(a)

$R' = SiMe_3, PH(NR_2)_2^+,$
$C_6H_3(CF_3)_2$

(b)

(c)

Figure 5.7 Isolable (a) Bertrand- and (b) Arduengo-type free carbenes predated by (c) transition metal complexes

In a similar manner, formaldehyde ($O=CH_2$) and parent imines $RN=CH_2$ are prone to oligomerization.

are generally more prone to subsequent reaction with co-ligands or external reagents. This may include other metal centres, leading to bridging carbene complexes (see later), but may be retarded by inclusion of bulky co-ligands.

The complications that occasionally arise in the use of diazoalkanes reflect the possible further reactions of carbene ligands, which will be dealt with subsequently, *e.g.* insertion into adjacent M–H or M–halide bonds and the formation of bimetallic complexes supported by bridging carbene ligands. In some cases, transition metals may catalyse reactions of diazoalkanes, leading to products which are suggestive of the reactions of free carbenes, *i.e.* dimerization, addition to alkenes (cyclopropanation) and insertion into C–H bonds (Figure 5.9). In such cases, however, the actual mechanism does not involve free carbenes but rather transient diazoalkane/carbene complexes. This is supported by the obser-

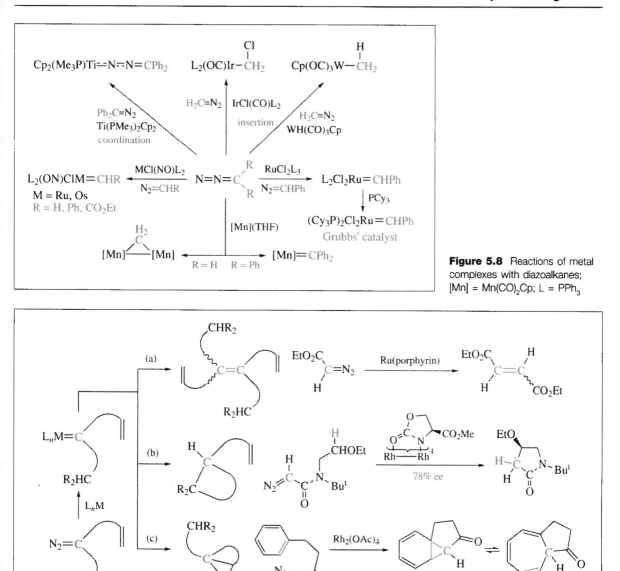

Figure 5.8 Reactions of metal complexes with diazoalkanes; [Mn] = Mn(CO)$_2$Cp; L = PPh$_3$

Figure 5.9 Transition metal catalysed (a) dimerization, (b) C–H insertion and (c) cyclopropanation reactions of diazoalkanes

vation that chiral metal catalysts can lead to enantioselective processes, implicating the *direct* participation of the optically active metal centre in the bond-forming step(s). The special case of cyclopropanation will be discussed below.

The search for halocarbenes led to Fischer's carbyne synthesis (see below); however, such halocarbene complexes are now widely available for elements of Group 6–9, albeit *via* alternative synthetic strategies. The first stable example arose from a study of the biochemically relevant reactions

Such a strategy led to the first complete series of chalcocarbonyl complexes OsCl₂(CA)(CO)(PPh₃)₂ (A = O, S, Se, Te) via reaction with H₂A/NaAH (Figure 3.31).

of halocarbons with porphyrin complexes under reducing conditions. Thus reaction of Fe(TPP)Cl with CCl_4 in the presence of iron powder (reductant) provided the diamagnetic dichlorocarbene complex Fe(TPP)(=CCl₂). Various other routes to halocarbene complexes are presented in Figure 5.10 (see also Figures 4.30b, 5.6). The key synthetic utility of these complexes lies in the ability of one or both halides to be replaced by a range of nucleophiles (*cf.* acyl halides and phosgene, O=CCl₂).

Figure 5.10 Synthetic routes to halocarbene complexes (see also Figure 4.30)

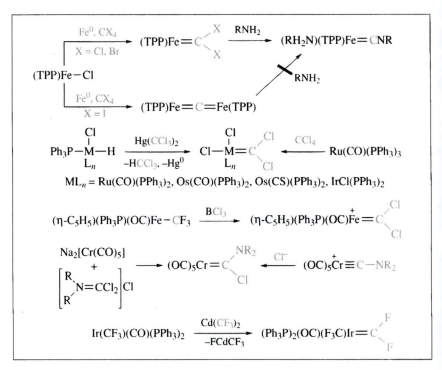

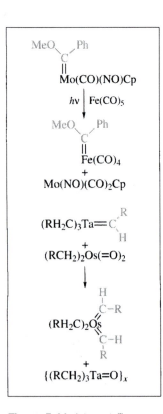

Figure 5.11 Intermetallic carbene transfer

Coordinatively unsaturated alkyl complexes devoid of β-protons are prone to α-elimination or abstraction reactions, especially for higher valent metals of Groups 5–7. Intramolecular α-elimination (H transferred to the metal) or α-abstraction (H transferred to a co-ligand) reactions (Figures 4.27–4.29) are complemented by reactions which result in removal of an α-C–H by an external reagent. Depending on the nature of the alkyl precursor, this hydrogen may be removed as either H⁻, H· or H⁺ (Figures 4.27). Many other alkyl α-substituents (halide, alkoxide, thiolate) are also capable of α-nucleofugic abstraction (Figure 4.30).

Alkylidene ligands may in some cases be transferred between metal centres; however, it should be stressed that these reactions do not involve the generation of free carbenes but rather bimetallic intermediates. So far, these reactions have involved transfer of a carbene ligand from an early transition metal to a later transition metal (Figure 5.11).

New carbene complexes may arise from the modification of a pre-formed carbene ligand. Examples have already been met in the case of halocarbenes (Figure 5.10); however, further examples will be discussed under the topic of carbene reactivity.

5.2.3 Reactions of Carbene Complexes

Both the HOMO and the LUMO of carbene complexes comprise contributions from both metal and carbon atomic orbitals. The relative contributions of these orbitals will be dependent upon the nature of the metal [charge, oxidation state, co-ligands, coordinative (un)saturation] and the carbene substituents (electronegativity, π-dative capacity). The effect of charge is illustrated by the two isoelectronic complexes $RM(=CH_2)Cp_2$ ($RM = MeTa, HW^+$). The methylene ligand in the tantalum complex (Figure 4.29) displays nucleophilic character, whilst the tungsten methylene is a strong electrophile (Figure 5.12).

Thermolysis

Many carbene complexes, especially those bearing one or two hydrogen substituents, are thermally unstable with respect to the formation of alkenes or their complexes (Figure 5.13), usually *via* bimolecular inter-mediates. For this reason the thermal stability of carbene complexes can normally be enhanced by inclusion of sterically demanding co-ligands. Often within a triad the stability of carbene complexes increases in the order 4d < 3d < 5d when analogous compounds can be obtained and compared.

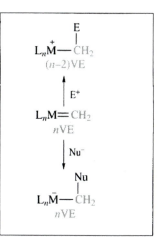

Figure 5.12 Metal-dependent methylene reactivity; *top*: L_nM = Ir(PMe₃)Cp*, Ta(CH₃)Cp₂, OsCl(NO)(PPh₃)₂, Irl(CO)(PPh₃)₂; *bottom*: L_nM = WHCp₂⁺, Fe(CO)₂Cp⁺, Mo(CO)₃Cp⁺, Os(dppm)Cp²⁺, Re(NO)(PPh₃)Cp⁺

Hydroxycarbenes present a special case wherein thermolysis usually affords the corresponding aldehyde.

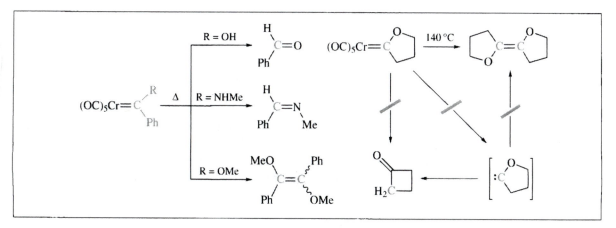

Ligand Substitution

Coordinatively unsaturated carbene complexes will normally bind further ligands to attain coordinative saturation (steric factors allowing).

Figure 5.13 Thermolysis of carbene complexes

For 18VE carbene complexes, however, direct attack by ligands at the metal centre is disfavoured. This leaves two mechanistic possibilities: (i) dissociative activation (Δ or *hv*), which may be enhanced by the characteristically high *trans* influence of carbene ligands; or (ii) associative activation *via* attack by the ligand (nucleophile) at the carbene carbon. Occasionally (*e.g.* Figure 5.14), both mechanisms may operate, the latter showing, as expected, a rate dependence on the nature of the incoming ligand (nucleophilicity, steric properties). Thus for phosphines of low nucleophilicities, rate limiting dissociation of a carbonyl co-ligand precedes rapid coordination of the phosphine. For smaller and highly basic phosphines, *e.g.* PMe₃ (Table 2.1), this route operates in addition to a second process involving nucleophilic attack at the carbene carbon (ylide formation, see below). This may be followed by either loss of a carbonyl ligand and migration of the phosphine to the metal (α-elimination), or alternatively competitive displacement of the ylide by a second phosphine.

Carbenes and carbynes can exert both a strong *trans* influence and *trans* effect. A **trans influence** is a ground-state destabilization (**thermodynamic**) of the *trans* M–L bond. The **trans effect** is a **kinetic** phenomenon wherein the lability of the *trans* ligand is increased either by ground-state destabilization or transition-state stabilization.

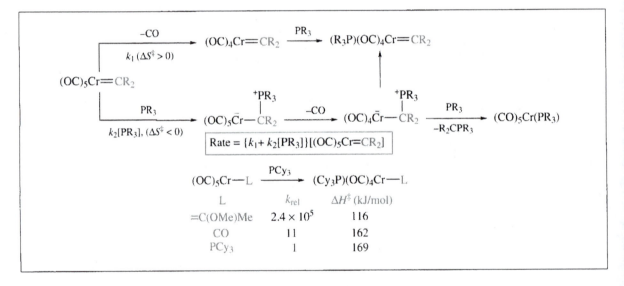

$$Rate = \{k_1 + k_2[PR_3]\}[(OC)_5Cr\!=\!CR_2]$$

L	k_{rel}	$\Delta H^{\ddagger}$ (kJ/mol)
=C(OMe)Me	2.4×10^5	116
CO	11	162
PCy₃	1	169

Figure 5.14 Dissociatively (k_1) and associatively (k_2) activated routes for ligand substitution; CR₂ = C(OMe)Me

In addition to carbenes exerting a strong *trans* influence (thermodynamic), which may be manifest as a *trans* effect (kinetic) in ligand substitution reactions, heteroatom-functionalized carbene ligands (with their associated dipolar resonance contributor, Figure 1.16) may assist in the stabilization of transition states of reduced coordination number by electron donation to the metal.

Nucleophilic Attack

For coordinatively unsaturated carbene complexes, nucleophilic attack may occur at the Lewis-acidic metal centre. For 18VE carbene complexes, however, nucleophiles generally attack at the carbene carbon. This process (Figures 5.12, 5.14) may be encouraged, as in the case of carbonyl ligands (Figure 3.20), by any factor which raises the energy of the carbon or metal orbitals contributing to the LUMO of the complex (positive charge, π-acidic co-ligands, electronegative carbene substituents). For simple Fischer-type carbenes, although the carbonyls carry a greater positive charge than the carbene carbon, nucleophiles generally attack at the carbene because of the high contribution that this atom makes to the LUMO (frontier orbital control) Figure 5.15 collects a range of reactions of Fischer-type carbenes with nucleophiles. For many monobasic nucleophiles (H–Nu), the ultimate product arises from substitution of alkoxide by Nu⁻. A zwitterionic intermediate (Figure 5.16) is assumed to occur and in some cases this intermediate may be isolated if the nucleophile is a tertiary amine (*e.g.* DABCO) or phosphine.

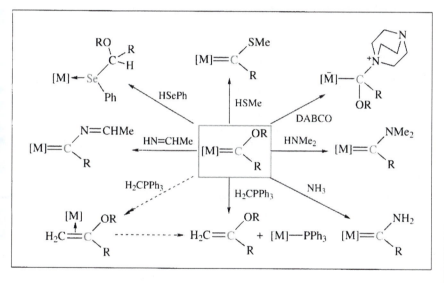

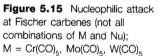

Figure 5.15 Nucleophilic attack at Fischer carbenes (not all combinations of M and Nu); M = Cr(CO)$_5$, Mo(CO)$_5$, W(CO)$_5$

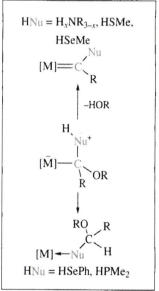

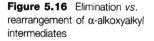

Figure 5.16 Elimination *vs.* rearrangement of α-alkoxyalkyl intermediates

Nucleophilic attack by hydride or organolithium reagents calls for special comment as it provides access to non-heteroatom-functionalized carbenes (alkylidenes). In both of these cases (Figure 5.17), an anionic α-alkoxyalkyl complex is formed and in some cases this may be isolated. However, more commonly, this intermediate is treated with a Brønsted acid at low temperature, resulting in alkoxide abstraction and formation of the alkylidene complex. The examples shown are somewhat thermally unstable, especially the benzylidene complex. These may, however, be generated *in situ* and trapped with a variety of reagents, *e.g.* the addition of chalcogens which provide complexes of chalcoaldehydes (E=CHR) and chalcoketones (E=CR$_2$). In some cases it is more convenient to prepare an α-alkoxyalkyl complex directly (Figure 4.30), prior to alkoxide abstraction.

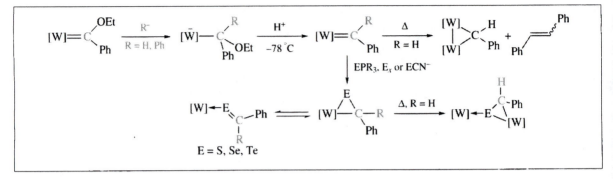

Figure 5.17 Benzylidene, diphenylcarbene and chalcoaldehyde complexes; [W] = W(CO)₅

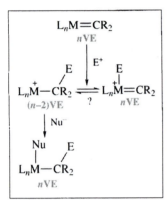

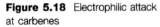

Figure 5.18 Electrophilic attack at carbenes

Electrophilic Attack

An electrophile may *ultimately* bind to either the metal centre (nVE) or the carbene carbon ($n - 2$VE), bearing in mind that the two potential outcomes (Figure 5.18) are related by the α-elimination–migratory insertion equilibrium (see also Figures 4.29, 5.12). More commonly, however, the conjugate nucleophile of the electrophilic reagent coordinates to the metal, blocking α-elimination, as illustrated for the electron-rich methylene complex Ir(=CH₂)(PMe₃)Cp* (Figure 5.19).

Intermediate situations may arise where the M=C bond may act as either an electrophile or a nucleophile (*i.e.* an amphiphile). This is illustrated by the complexes Ru(=CF₂)(CO)₂(PPh₃)₂ [Ru(0), d⁸] and Re(=CHR)(CO)₂Cp [Re(I), d⁶] (Figure 5.20). The Ru(0) example is particularly illustrative in that the related Ru(II) complex, RuCl₂(CO)-(=CF₂)(PPh₃)₂ (d⁶), reacts solely with nucleophiles. The intermediate character of these and other amphiphilic carbenes serve to devalue the

Figure 5.19 Electrophilic attack at an electron-rich carbene; [Ir] = Ir(PMe₃)Cp*; X = Cl (as [PhNH₃]Cl), H, OC₆H₄CF₃, SBuᵗ, N(CO)₂C₂H₄

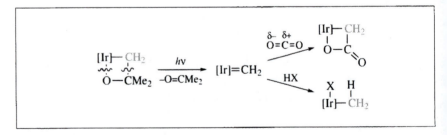

Figure 5.20 Amphiphilic carbenes; [Ru] = Ru(CO)₂(PPh₃)₂; [Re] = Re(CO)₂Cp; R = CH₂CH₂Buᵗ

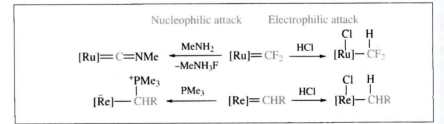

Fischer/Schrock dichotomy in favour of a more inclusive and tunable reactivity continuum.

Cyclopropanation and Related Reactions

Free carbenes readily add to C–C multiple bonds with the formation of cyclopropanes. Cyclopropanes may also be obtained *via* the reaction of carbene complexes with alkenes; however, since free carbenes are not involved, the course of the reaction may be influenced by the nature of the metal complex. Generally (though not exclusively), the reaction is considered to proceed *via* coordination of the alkene *cis* to the carbene ligand, followed by a [2 + 2] cycloaddition of the C=C and M=C double bonds. The resulting metallacyclobutane may then undergo reductive elimination of the cyclopropane (Figure 5.21a). For particularly electrophilic (18VE) carbene complexes, however, an alternative mechanism may operate, in which the carbene carbon attacks the alkene directly to generate a γ-carbocationic alkyl ligand which collapses with ring closure and electrophilic alkyl cleavage (Figure 4.24). Support for this comes from unequivocal synthesis of a putative intermediate *via* reaction of γ-haloalkyl complexes with halide abstractors (Figure 5.21b).

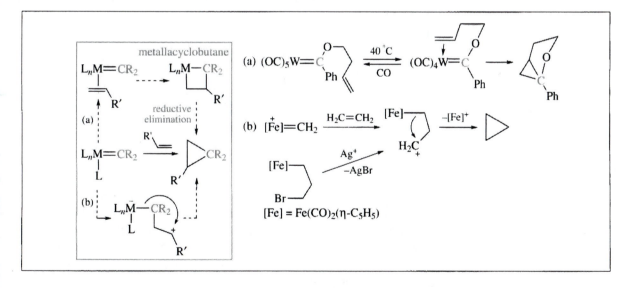

This stoichiometric process may be made catalytic if a means of reintroducing a carbene ligand can be incorporated into a cycle. This has been achieved now for a number of metal catalysts, with diazoalkanes typically serving as the carbene source (Figure 5.9c). The key intermediate in the above process is a metallacyclobutane, to which we will return (see Alkene metathesis below). First, however, we will consider the reactions of metal carbenes with other C–element double bonds, *i.e.*

Figure 5.21 Alkene cyclopropanation *via* (a) cycloaddition or (b) electrophilic attack processes

This general phenomenon is particularly useful for some classes of deactivated organic carbonyl compound, e.g. esters and amides, which do not undergo Wittig alkenation.

C=O and C=N. In the reactions of aldehydes with phosphorus ylides ('phosphorus alkylidenes'), there is a strong impetus for the formation of phosphine oxide and alkene via a four-membered intermediate: the **Wittig reaction**. For transition metal alkylidenes, a similar situation arises: those of the earlier electropositive (i.e. hard, oxophilic) elements react with organic carbonyl complexes via reactions analogous to the Wittig reaction, resulting in the formation of the corresponding metal oxo complex and liberation of an alkene (Figure 5.22).

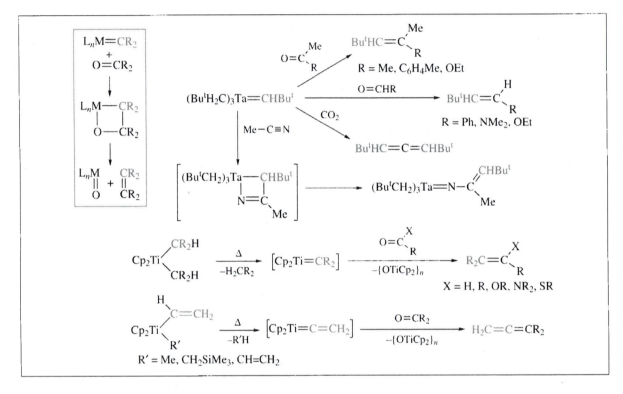

Figure 5.22 Alkylidene cyclo-additions with heteroalkenes

Many examples of this class of reaction have been demonstrated for isolable alkylidene complexes, in addition to similar results obtained for complexes which appear to react as alkylidenes, e.g. the Petasis reagent $TiMe_2Cp_2$ and Tebbe's reagent, $TiCl(CH_2AlMe_2Cl)Cp_2$ (Figure 4.14), both of which alkenate esters presumably via a reactive methylene complex, $Ti(=CH_2)Cp_2$. Imines also react with some early transition metal alkylidene complexes to provide the corresponding alkene and imido complex.

Alkene Metathesis (Dismutation)

Returning to the key metallacyclobutane intermediate above, we must now consider an alternative decomposition mechanism which may operate. The reverse of the [2 + 2] cycloaddition would regenerate the original alkylidene-alkene complex, *if the complex had a 'memory'*. However, once formed, the metallacyclobutane may break down in two ways: from whence it came or alternatively to generate new alkylidene and (labile) alkene ligands. Many factors, both thermodynamic and kinetic, may determine which of these is favoured. Since the new alkene can dissociate from the metal centre and be replaced by a further molecule of the original alkene substrate, a process arises by which the halves of an alkene may be exchanged or 'metathesized'. This alkene metathesis process in its various forms underpins a range of technologically useful alkene transformations. The accepted (Chauvin–Herisson) mechanism (Figure 5.23) is supported by the observation that the process may be initiated by either preformed alkylidene or metallacyclobutane precatalysts. Variants will be considered in turn, although the same mechanism accounts for each, the distinctions arising from the nature of the substrates and products. As is often the case, applications of the process predated its understanding. Various effective heterogeneous catalysts have long been known and generally arise from a high-valent transition metal compound (Re_2O_7, WCl_6, $MoCl_5$) in combination with a nucleophilic alkylating agent (SnR_4, AlR_3, ZnR_2). Such systems are believed to generate polyalkyls which convert by α-H abstraction/elimination reactions to alkylidene species. The true nature of the active site(s), however, often remains obscure, the inevitable downside of heterogeneous catalysis.

In contrast to ill-defined heterogeneous systems, well-defined catalytically active molecular complexes of the metallacyclobutane or alkylidene variety have each been employed as initiators. The requirement for initial alkene coordination is reflected in the observation that the most effective catalysts tend to be four-coordinate alkylidene complexes. Less active catalysts may be activated by co-ligand removal, which may be effected by Lewis acids (Al_2Cl_6, Ga_2Cl_6, CuCl) capable of abstracting halide or phosphine ligands. Thus W(=O)(=CHBu^t)Cl$_2$(PEt$_3$)$_2$ is activated by Al_2Cl_6 (halide abstraction) whilst CuCl abstracts phosphine from Grubbs' catalyst, RuCl$_2$(=CHPh)(PCy$_3$)$_2$. The tendency of early transition metal alkylidenes to form either oxo or imido complexes on reaction with carbonyl or imine functional groups complicates the design of alkene metathesis catalysts based on these metals, since they are somewhat intolerant of many polar functional groups (requiring protection/deprotection protocols). Alkylidene complexes of the later transition metals, in particular Grubbs' catalyst (Figure 5.8), are less oxophilic and can tolerate alkenes bearing a range of oxygen- and nitrogen-based polar functional groups. In general, however, the heteroatom must be remote from the double bond to be metathesized since the inclusion of heteroatoms directly on the double bond would result in the formation

Obscure pre-catalysts include $Re_2O_7/Al_2O_3/SnMe_4$, $WCl_6/EtOH/Al_2Et_2Cl_4$, $WCl_6/SnMe_4$, $WOCl_4/Al_2Et_2Cl_4$, $MoCl_2(NO)_2(PPh_3)_2/Al_2Me_3Cl_3$ and $(OC)_5W=C(OMe)Ph/SiCl_4/SnCl_4$.

Not all pre-catalysts require an alkylating agent; in some cases, alkenes are capable of rearrangement to a transient alkylidene *via* **hydrogen shifts** upon coordination.

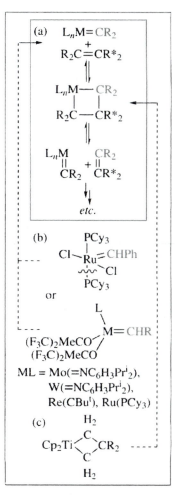

Figure 5.23 (a) Chauvin–Herisson mechanism for alkene metathesis with well-defined (b) alkylidene and (c) metallacyclobutane initiators

of heteroatom-functionalized carbene complexes. These usually only show limited activity relative to non-heteroatom carbene (alkylidene) complexes.

Acyclic Diene Metathesis (ADMET)

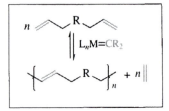

Figure 5.24 ADMET polymerization

ADMET is applied to the synthesis of polymers from dienes, wherein a sacrificial alkene (most commonly ethene) is lost as the polymerization proceeds. The strength of this process (and ROMP, below) is that polymers with regularly repeating alkene groups (polyenamers) are obtained with a regularity, and proximity to other repeating functional groups, strictly defined by the nature of the diene monomer. The process is, however, in principle an equilibrium, and accordingly by varying the conditions such polymers may be depolymerized in the presence of a catalyst and excess ethene (Figure 5.24).

Ring-opening Metathesis Polymerization (ROMP)

Figure 5.25 ROMP: mechanism and application

For cyclic alkenes which involve a degree of ring strain, this may be eliminated by metathesizing the halves of the strained alkene, resulting in linear polyalkenes (polyenamers) (Figure 5.25). This is illustrated by the ROMP of norbornene to provide poly(norbornene). The Diels–Alder

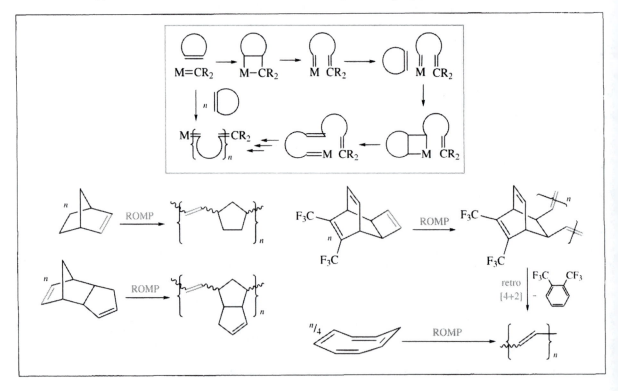

dimer dicyclopentadiene (a substituted norbornene) comprises two cyclopentene rings, either of which might in principle be opened. However, the more strained of these which is involved in the bicyclic ring system is preferentially opened, although highly active catalysts may in the later stages of catalysis also open the second ring, leading to polymer cross-linking. Cyclobutenes also involve considerable ring strain and the ROMP of one example provides access to the conducting polymer precursor polyacetylene, $(CH=CH)_n$.

Polyacetylene can also be obtained from the ROMP of cyclooctatetraene by a titanacyclobutane.

Ring Closure Metathesis

The difference in enthalpies (*e.g.* ring strain) of substrate and product provides a driving force for the ROMP process. Entropic considerations may also play a part, as in the case of ring closure metathesis (RCM) of α,ω-dienes (Figure 5.26). With the ultimate generation of two product molecules from one substrate molecule, and so long as the resulting cycloalkene does not possess prohibitive ring strain, the gain in entropy $(\Delta S > 0)$ drives the reaction to completion. If the liberated sacrificial alkene is volatile (ethene, butene), removal may further displace the

Figure 5.26 Ring closure metathesis including enantioselective applications

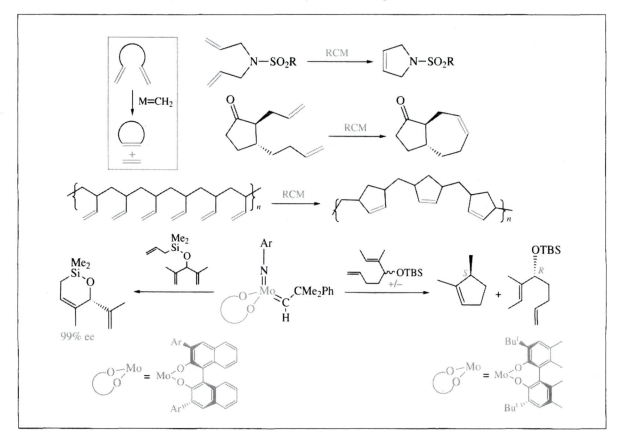

equilibrium in favour of cycloalkene isolation. Entropic factors may, however, count against the formation of larger rings. Furthermore, as these have less ring strain, the double bond that results may have E or Z regiochemistry (Figure 5.26). Catalytic asymmetric ring closure metathesis (ARCM) has also been developed in various enantioselective forms. Kinetic enantiodifferentiation may be achieved such that one enantiomer of a racemate may be selectively closed, ideally leaving the other untouched. Alternatively, a prochiral substrate with three alkene groups, two of which are diastereotopically related, may undergo ARCM with a kinetic preference for one. In these processes, the optical induction is critically dependent on the nature of the catalyst and chiral co-ligands (Figure 5.26)

Ligand Coupling Reactions

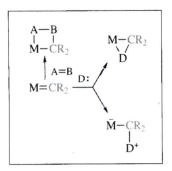

Figure 5.27 Carbene coupling with 2VE ligands

The coupling of carbene ligands with alkenes was discussed above; however, many other ligands may also enter into coupling reactions with carbenes. Simple 1VE ligands (hydrides, alkyls) may migrate onto a carbene ligand to generate an alkyl, the former being the microscopic reverse of α-M–H elimination (Figures 4.25, 4.27). These reactions deplete the valence electron count by 2VE and accordingly may be most readily observed when a new ligand is added to saturate the VE count for the resulting alkyl complex. By geometric necessity, the migrating ligand must first occupy a coordination site *cis* to the carbene. Various 2VE ligands (including alkenes discussed above) have also been shown to couple with carbenes (Figures 3.23, 5.27, 5.28), in processes that in the absence of added ligands also lead to a loss of 2VE. The special case of alkynes is discussed next.

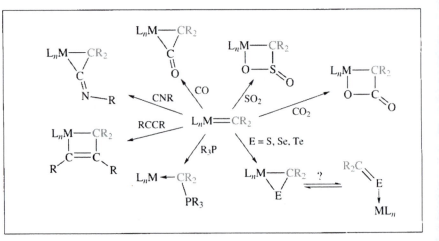

Figure 5.28 Carbene coupling with 2VE ligands: examples

The Dötz and Related Reactions

Alkynes as well as alkenes can enter into cycloaddition reactions with carbene ligands. The initially formed metallacycle is a metallacyclobutene, *e.g.* in the reaction of Tebbe's reagent with diphenylacetylene (Figure 5.29). One of the most versatile stoichiometric applications of carbene complexes, the Dötz reaction (Figure 5.30), is based on this process applied (primarily) to chromium carbenes. Generally the intermediate metallacyclobutene (or its vinylcarbene isomer) is not usually isolated. Rather, subsequent combination of this ligand with a carbonyl co-ligand ultimately leads to the formation of annulated species, apparently *via* formation of an electrophilic vinylketene complex (Figures 3.23, 5.28).

For aryl carbenes the benzannulated organic product may remain coordinated to the chromium in a *hexahapto* manner (Chapter 7), or be liberated *via* mild oxidation (Ce^{IV}, Me$_3$NO) or carbonylation [–Cr(CO)$_6$].

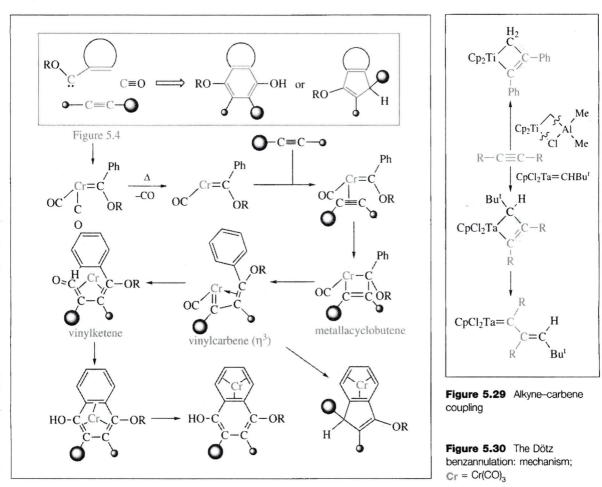

Figure 5.4

vinylketene

vinylcarbene (η^3)

metallacyclobutene

Figure 5.29 Alkyne–carbene coupling

Figure 5.30 The Dötz benzannulation: mechanism; Cr = Cr(CO)$_3$

β-Deprotonation

Just as the β-protons of acyl ligands are acidic, allowing the synthesis of β-functionalized acyl ligands (Figure 4.17), the β-protons of many carbene ligands may be removed by non-nucleophilic bases. Nucleophilic bases are generally avoided since these may directly attack the carbene carbon. The resulting anionic carbene complexes may then be functionalized with a variety of electrophiles (Figure 5.31).

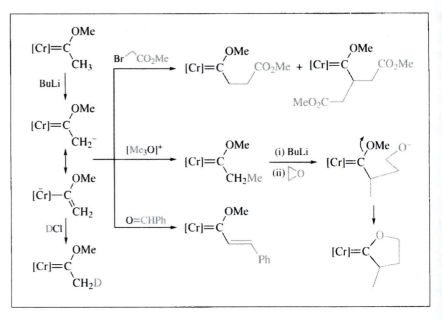

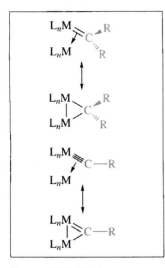

Figure 5.31 Carbene β-H acidity: synthetic applications; Cr = Cr(CO)$_5$

Figure 5.32 Valence (resonance) contributors for μ-carbenes and μ-carbynes

5.2.4 Bridging Carbenes

The loose analogy between M=C and C=C bonds has been drawn above, whilst the ability of alkenes to coordinate to a metal centre has also been met in passing, prior to a fuller treatment in Chapter 6. It is therefore reasonable to now consider the possibility that M=C double bonds might coordinate to a metal centre. In fact this possibility has already been seen, implicitly in the formation of Mn$_2$(μ-CH$_2$)(CO)$_4$Cp$_2$ from diazomethane and Mn(THF)(CO)$_2$Cp, *via* the putative intermediate complex Mn(=CH$_2$)(CO)$_2$Cp which 'coordinates' its Mn=C double bond to a further Mn(CO)$_2$Cp fragment. This represents the history of the process; however, once formed, we might view the product as a bridging carbene complex, a dimetallacyclopropane or a metal complex of Mn(=CH$_2$)(CO)$_2$Cp (Figure 5.32).

This last view leads to the principles developed by Stone, which allow the planned synthesis of a wide range of complexes involving two (or more) different metal centres (heterobimetallics) bridged by carbene ligands (Figure 5.33). Many other bi- and trimetallic carbene complexes may, however, also be prepared by beginning with the preformed bi- or trimetal core. The chemistry of bridging carbenes will not be dealt with specifically in this text other than to point out that carbenes are believed to occur on the surface of metals and metal oxides in some of the heterogeneously catalysed transformations of organic substrates (*e.g.* the Fischer–Tropsch process). These molecular bimetallic carbene complexes serve as models for such species (the cluster–surface analogy, Figure 2.19). Given that the 'carbene' carbon is effectively sp^3 hybridized in such species, the description as dimetallacyclopropanes has some merit.

5.3 Carbynes (Alkylidynes)

Alkylidyne or carbyne complexes involve formal metal–carbon triple bonds. The bonding in these complexes is most simply envisaged by considering the frontier orbitals of a cationic carbyne fragment [CR]$^+$, in combination with a metal centre with occupied orbitals of π-symmetry with respect to the M–C vector (Figure 5.34). The interactions can then be seen as analogous to those of a carbonyl ligand (or [NO]$^+$), *i.e.* combination of a dative interaction of σ-symmetry coupled synergically with two (orthogonal) retrodative (π-acid) interactions.

Given the topological parallels between Figures 1.12 and 5.34, it can be expected that the arguments presented for the reactivity of carbene complexes towards nucleophiles and electrophiles could be equally applicable to carbyne complexes, which in general they are. Indeed, this included premature attempts to dichotomize carbyne complexes within 'Fischer-type' and 'Schrock-type' regimes of reactivity. Just as heteroatom substitutents compromise the M=C bond of carbenes, carbyne ligands which bear heteroatom substituents (*e.g.* Figure 3.19) tend to show longer and less reactive M=C bonds (Figure 5.35).

5.3.1 Characterization of Carbyne Complexes

As in the case of carbene complexes, ^{13}C NMR spectroscopy is particularly useful in that the carbyne carbon typically resonates to low field (240 and 360 ppm), with heteroatom substituents shifting this to higher field. As noted above for carbene complexes, X-ray crystallography reveals that carbyne complexes have very short metal–carbon bonds, typically the shortest of any metal–carbon multiple bond, but lengthened if heteroatom substituents are present.

5.3.2 Synthesis of Carbyne Complexes

Free carbynes, 'CR', are exceedingly reactive short-lived species and

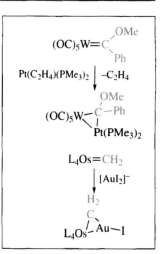

Figure 5.33 Carbene bridge-assisted metal–metal bond formation; OsL_4 = OsCl(NO)(PPh$_3$)$_2$

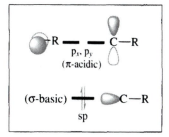

Figure 5.34 Carbyne [CR]$^+$ frontier orbitals for synergic bonding with transition metals

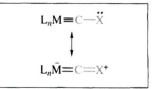

Figure 5.35 Heteroatom-substituted carbynes; X = NR$_2$, OR, SR, SeR, TeR

therefore entirely inappropriate precursors for carbyne complexes. Furthermore, their limited chemistry finds no analogy in the organometallic chemistry of carbyne complexes. As with carbene chemistry, the genesis of carbyne complexes occurred in the laboratory of E. O. Fischer. Fischer's alkoxide abstraction route (Figure 5.2) has been extended to all the metals of Groups 6 and 7 as well as iron. Furthermore, refinements have made it unnecessary in many cases to isolate the intermediate carbene or acylate complexes. Rather, the acyl metallate intermediates may be often converted directly to carbynes by treatment with oxide (O²⁻) abstracting reagents (oxalyl halides, phosphine halides, thionyl chloride, phosgene, triflic or trifluoroacetic anhydrides) (Figure 5.36). This approach relies on converting the oxide into a better leaving group than an alkoxide. This is illustrated by the reaction of acyl metallates with acetic or trifluoroacetic anhydride. With the former, a stable (though synthetically versatile) acetoxycarbene is formed. However, with the latter, rearrangement to the trifluoroacetatocarbyne complex occurs spontaneously.

$CF_3CO_2^-$ is a better leaving group than $CH_3CO_2^-$.

The stability may be increased by employing amino (**thermodynamic stabilization**) or sterically demanding (**kinetic stabilization**) carbyne substituents.

Figure 5.36 Carbyne synthesis *via* oxide abstraction; A–X = ClCOCO–Cl, ClCO–Cl, CF_3CO–O_2CCF_3, $BrPh_3P$–Br, ClOS–Cl, F_3CSO_2–O_3SCF_3)

Fischer's carbyne complexes are thermally labile (facile loss of CO) and decompose bimolecularly with formation of alkynes. This carbonyl lability may be traced to the strong π-acidity of the carbyne ligand, which compromises retrodonation to the carbonyl ligands. On treatment with donor ligands, however, thermally stable substituted derivatives may be obtained (Figure 5.36).

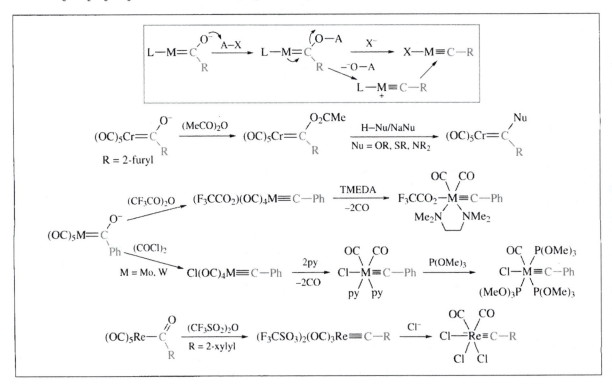

Just as alkyl ligands may undergo α-hydrogen abstraction or elimination to provide alkylidene ligands, these in turn may be prone to α-
hydrogen elimination or abstraction to generate alkylidyne ligands. These
reactions are most prevalent for metals of Groups 5–7 but have occasionally been observed within Group 8, a trend already noted for alkyl/
alkylidene chemistry. Schrock's archetypal tantalum neopentylidene
complex (Figure 4.28) may be deprotonated by strong bases to provide
an anionic carbyne complex (Figure 5.37). Alternatively, reduction of
related complexes may be accompanied by α-hydrogen elimination.
Further examples of alkylidyne syntheses *via* this approach are shown
in Figure 5.37. Amongst these, the complexes $W(\equiv CR)(CH_2R)_3$ and
$Os(=CHR)_2(CH_2R)_2$ present an interesting pair in that the presence of
an extra 2VE for osmium appears to favour the bis(carbene) tautomer
(in the ground state). Since this compound also results from the addition
of $LiCD_2R$ to $Os(\equiv CR)(CH_2R)_2(OTf)(py)_2$, with deuterium scrambled
between alkyl and alkylidene sites, it may be surmised that this species
may well be in equilibrium with an undetected alkylidyne tautomer,
$Os(\equiv CR)(CH_2R)_3$.

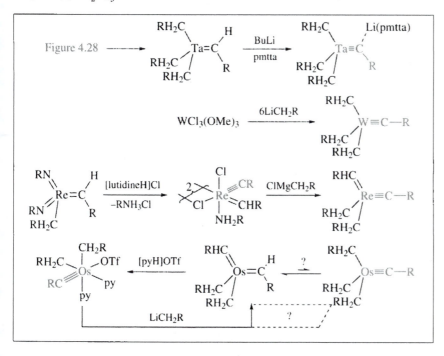

Figure 5.37 Carbyne syntheses
via α-H elimination/abstraction

Electrophilic attack at the β-atom of CO, isocyanides and chalcocarbonyls has already been met as a route to carbyne complexes (Figure
3.19). This reaction may be extended to the β-carbon of vinylidenes,
which are discussed below. The remaining routes to carbyne complexes
at present lack generality; however, selected examples are included in

Figure 5.38. In recent times these have included the rearrangement of terminal alkynes in combination with late transition metal hydrido complexes, presumably *via* vinylidene intermediates (see below).

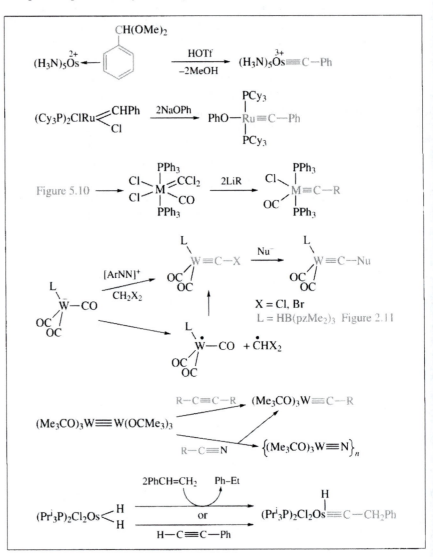

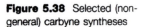
Figure 5.38 Selected (non-general) carbyne syntheses

5.3.3 Reactions of Carbyne Complexes

The reactions of carbyne complexes will be discussed, albeit more briefly, along the lines developed for alkylidene complexes, to emphasize the parallels. Once again, as with alkylidenes, early generalizations suggested two classes of complexes, Fischer-type and Schrock-type, which have subsequently been recognized as extremes in a tunable reactivity continuum. Furthermore, within the chemistry of tungsten, Fischer-type carbynes may be converted to Schrock-type by simple oxidation, or *vice versa* by reduction. As with carbenes, the reactivity of the metal–carbon triple bond is crucially dependent upon the nature of the metal centre [charge, oxidation state, co-ligands, coordinative (un)saturation] and the carbyne substituent (*e.g.* π-dative capacity).

Electrophilic Attack

Electrophilic attack at carbyne complexes may *ultimately* place the electrophile on either the metal or the (former) carbyne carbon, the two possibilities being related in principle by α-elimination/migratory insertion processes (Figure 5.39). The reactions of the osmium carbyne complex are suggestive of an analogy with alkynes. Each of these reactions (hydrohalogenation, chlorination, chalcogen addition, metal complexation; see below) have parallels in the chemistry of alkynes.

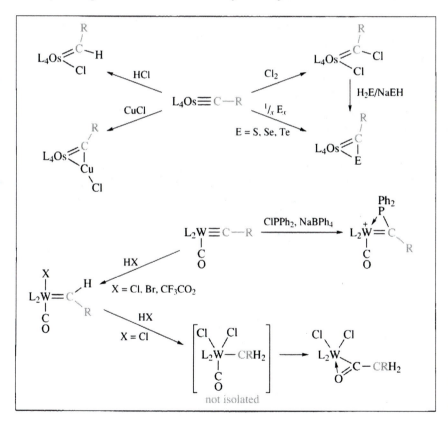

Figure 5.39 Electrophilic attack at carbyne ligands; OsL$_4$ = OsCl(CO)(PPh$_3$)$_2$; WL$_2$ = W(CO)Cp

Nucleophilic Attack

An example of this class of reactivity has already been met in the formation of halocarbenes *via* the reaction of cationic carbyne complexes

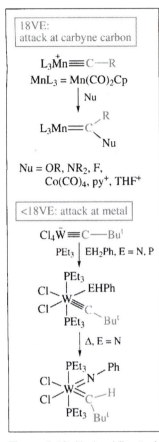

Figure 5.40 Nucleophilic attack at carbyne ligands

Figure 5.41 Migratory insertion of carbyne and hydride ligands; $WL_3 = W(CO)\{P(OMe)_3\}_2$; R = 2,4, 6-$Me_3C_6H_2$; $OsL_4 = OsCl_2(PPr^i_3)_2$

with halides (Figure 5.10), and alluded to as a synthetic route to carbene complexes in general (Figure 5.6). This type of reaction will be particular favoured for cationic 18VE complexes and/or those with strongly π-acidic co-ligands. Further examples are shown in Figure 5.40, including one in which the nucleophile is a carbonyl metallate, leading to a binuclear bridging carbyne complex. Note that for coordinatively unsaturated complexes the metal centre presents an alternative site for nucleophilic attack. In the case of monobasic nucleophiles, this may be followed by transfer of a hydrogen atom to the alkylidyne carbon and formation of a carbene ligand.

Since carbynes may arise from α-elimination reactions, it follows that the reverse, migratory insertion, is also possible. This is illustrated for the case of hydrido-carbyne complexes (Figure 5.41)

Ketenyl Formation

Figure 3.23 introduced the possibility of carbonyl ligands coupling with either alkylidene or alkylidyne ligands. The process depletes the VE count by two units and accordingly is most readily observed when an external ligand is added to trap, in the case of alkylidynes, the resulting 3VE ketenyl ligand. The ketenyl ligand may, however, in the presence of excess and strongly nucleophilic ligands assume a *monohapto* (1VE) coordination. This type of reactivity is characteristic of alkylidynes of Groups 6 and 7 and may also be induced photochemically. Isocyanide ligands may also couple with carbynes to provide iminoketenyls. The reverse reaction has also been observed. Of the various canonical forms which may describe ketenyl ligands, dipolar contributions ('oxyalkyne') suggest a nucleophilic character for the ketenyl oxygen which is indeed reflected in reactions with various electrophiles (Figure 5.42).

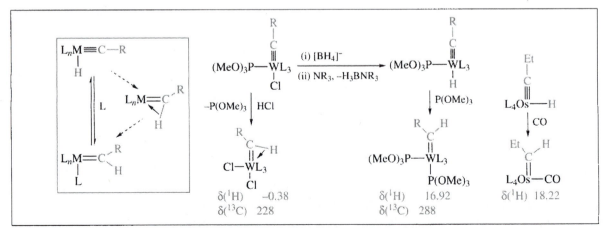

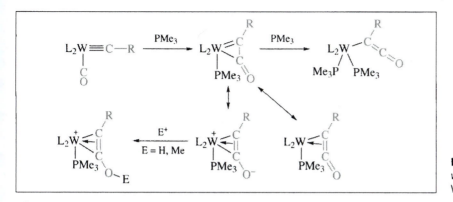

Figure 5.42 Ketenyl formation *via* carbyne–carbonyl coupling; WL_2 = W(CO)Cp

Cycloaddition Reactions

Alkynes enter into many cycloaddition reactions and by analogy so do some alkylidyne complexes. Selected examples are shown in Figure 5.43; however, those involving alkynes are reserved for discussion in the following section (alkyne metathesis).

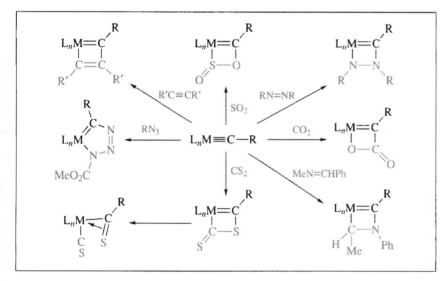

Figure 5.43 Selected cycloaddition reactions of carbynes

Alkyne Metathesis

By analogy with alkene metathesis, carbyne complexes might be expected to mediate the metathesis of alkynes, which indeed they do, but with some specific limitations. The basic mechanism parallels that for alkene metathesis, with the key intermediate being a metallacyclobutadiene which may break down in one of two possible directions (Figure 5.44).

Two complications however arise: (i) the metallacyclobutadiene may convert to a η^3-cyclopropenyl complex (see Figures 7.18, 7.19); (ii) for

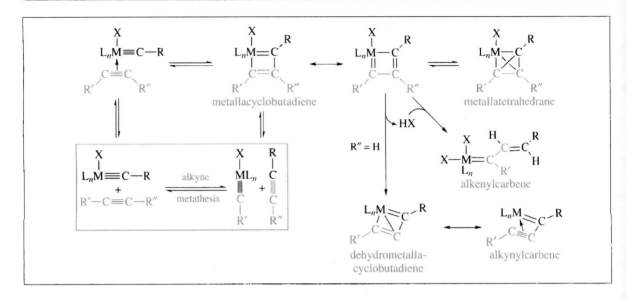

Figure 5.44
Metallacyclobutadienes and
dehydrometallacyclobutadienes:
relevance to alkyne metathesis
(see also Figure 7.18)

metallacyclobutadienes derived from terminal alkynes, deprotonation may occur to provide catalytically inactive dehydrometallacyclobutadienes. Active alkyne metathesis catalysts (both obscure and well-defined examples) are generally limited to early transition metals, with the attendant intolerance of polar functional groups already encountered for alkene metathesis processes. However, this reflects the comparative immaturity of the field with respect to the technologically more useful and pursued area of alkene metathesis. It may be anticipated that more sophisticated and versatile alkyne metathesis catalysts will arise. The reaction of $W_2(OR)_6$ with alkynes or nitriles provides alkylidyne complexes which may used *in situ* for alkyne metathesis, or the synthesis of new carbyne complexes which are not easily available via the conventional routes (Figure 5.45). In these preparations, substituted propynes are employed so that the equilibrium may be driven in the direction of the desired carbyne complex by the loss of volatile but-2-yne from the reaction mixture. This is illustrated for the synthesis of a polymeric carbyne complex from 4-propynylpyridine, the metathesis polymerization of 1,4-bis(propynyl)benzene to provide but-2-yne and a conjugated poly-ynamer, and an example of ring-closure alkyne metathesis (Figure 5.45).

5.3.4 Bridging Carbyne Complexes

Carbyne ligands may bridge two (μ) or three (μ_3) metal centres, providing a total of 3VE to the overall electron count. Earlier synthetic routes to such complexes involved the reactions of carbonyl metallates with 1,1,1-trihaloalkanes, or the cleavage of alkyne ligands coordinated to clusters (Figure 5.46).

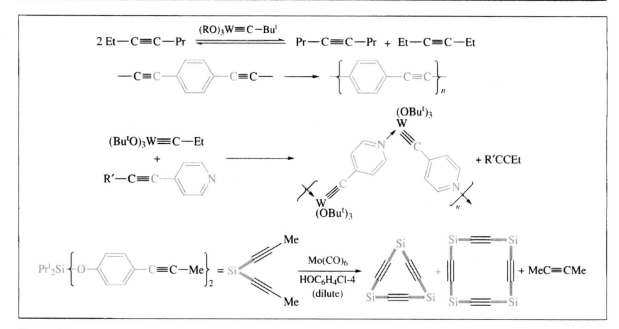

Figure 5.45 Alkyne metathesis: applications

Figure 5.46 Bridging carbyne complexes; Co = Co(CO)$_3$; Ru = Ru(CO)$_3$

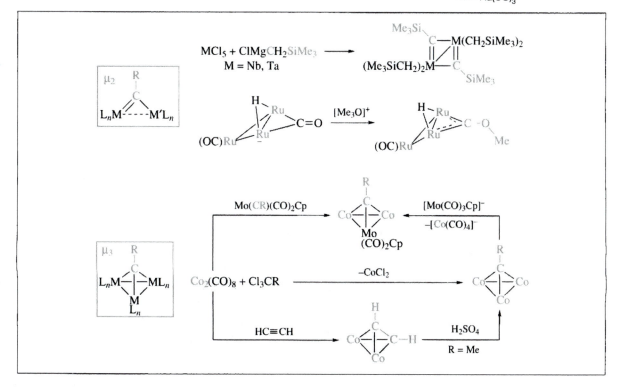

An alternative strategy considers the isoelectronic analogy between M≡C and C≡C triple bonds. If an alkyne can coordinate to a metal centre (Chapter 6), then in principle so can a M≡C bond (Figure 5.32). This approach, pioneered again by Stone, has proven particularly versatile for the designed synthesis of bi- and polymetallic complexes bridged by one or more alkylidyne units. The application of this approach is illustrated for the alkylidyne complex W(≡CR)(CO)$_2$Cp in Figure 5.47. The description of such complexes as dimetallacyclopropenes suggests that one M=C double bond remains, and accordingly a second metal may be added to this. Alternatively, bimetallic complexes may add directly to the M≡C bond to give triply bridging trimetallatetrahedranes. Finally, two alkylidyne complexes may be coupled by a further metal centre, in a manner reminiscent of alkyne coupling (Chapter 6), further strengthening the analogy.

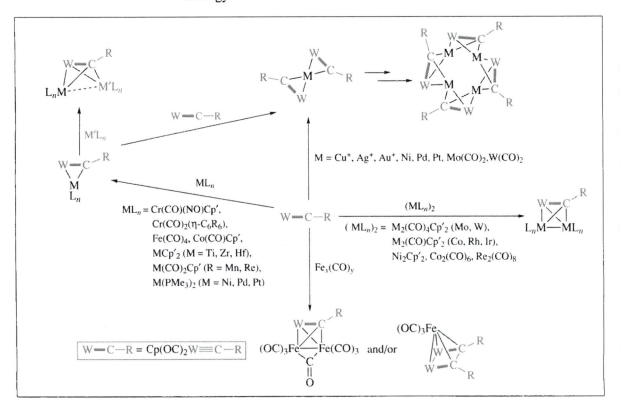

Figure 5.47 Stone alkylidene condensations

Vinylidenes, C=CHR, have been generated in the gas phase, but rearrange extremely rapidly to the alkynes, HC≡CR.

5.4 Vinylidenes (Alkenylidenes)

If there is an analogy between alkenes/alkynes and alkylidenes/alkylidynes, then the metal-based analogue of an allene would be a vinylidene (alkenylidene), L$_n$M=C=CR$_2$, with cumulenated M=C and C=C double bonds. Such complexes are well established and generally arise directly or indirectly from transformations of coordinated alkynes (Figures 4.12, 4.18). Complexes of alkynes bearing H, SiR_3, SnR_3, SR and SeR groups are prone to 1,2-shifts of these substituents to generate vinylidene ligands (Figure 5.48).

An alternative and widely employed approach involves the initial synthesis of a σ-alkynyl complex, followed by treatment with a variety of electrophiles. As with the heterocarbonyls shown in Figure 3.19, the atom β to the metal is activated towards electrophilic attack (Figure 5.49). This strategy is most effective for alkynyls bound to highly π-basic (commonly d^6) metal centres for two reasons. Firstly, the more electron rich the metal centre, the more the β-carbon will be activated (via retrodonation) to electrophilic attack. Secondly, the vinylidene ligand is a potent π-acid, and will accordingly be most stabilized by coordination to a π-basic metal centre.

The π-acidity of vinylidene ligands also points towards reactions with nucleophiles at C_α. This will be particularly true if the metal centre is positively charged or co-ligated by other strong π-acids, which competitively compromise $M \rightarrow C_\alpha$ retrodonation. In the case of monobasic

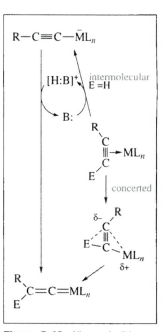

Figure 5.48 Alkyne–vinylidene rearrangements; E = H, SiR_3, SnR_3, SR, SeR

Figure 5.49 Vinylidene synthesis: (a) electrophilic attack at alkynyl ligands, (b) alkenyl α-M–H elimination and (c) acyl dehydroxylation (see also Figure 4.18); [Ru] = $Ru(PPh_3)_2Cp$; [Fe] = $Fe(CO)(PPh_3)Cp$; Tf = SO_2CF_3

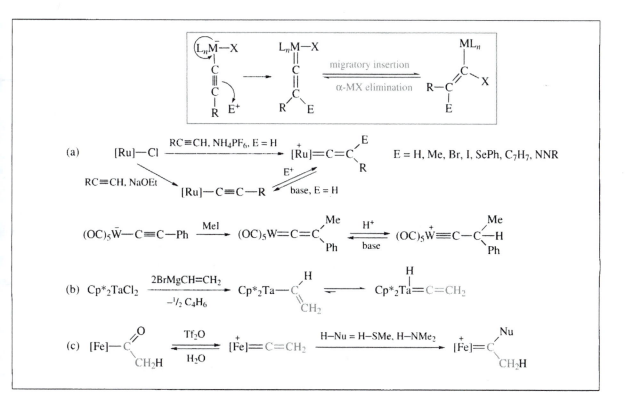

nucleophiles (H–Nu), the resulting vinyl ligands may be prone to protonation at C_β, resulting in the formation of a carbene ligand (Figures 4.18, 5.49). Two special cases of this reaction manifold deserve comment. (i) The reaction of 4-hydroxyalk-1-ynes with various transition metal centres may lead directly to cyclic carbene ligands *via* internal nucleophilic attack by the alcohol at the C_α of a vinylidene intermediate (Figure 5.6). (ii) By using water as the nucleophile, a hydroxycarbene may be formed which rearranges to an acyl ligand. If α-elimination (the reverse of migratory insertion) of the acyl substituent occurs, a complex results, wherein a carbonyl and an alkyl ligand arise from cleavage of the C≡C triple bond (!).

Just as alkylidenes and alkylidynes may bridge M–M bonds, vinylidenes may also be employed to support metal–metal bonds. This approach has included the addition of preformed terminal vinylidene complexes to unsaturated metal complexes or, occasionally, the direct assembly from terminal alkynes with unsaturated metal complex precursors (Figure 5.50).

Figure 5.50 Bridging vinylidenes; [Mn] = Mn(CO)$_2$Cp

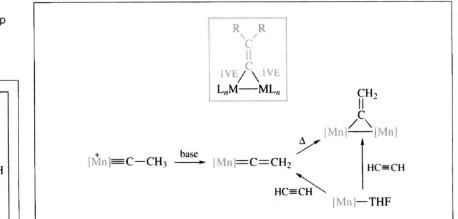

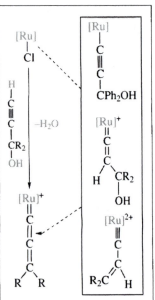

Figure 5.51 Allenylidene synthesis: propynol dehydration; [Ru] = RuCl(dppm)$_2$, Ru(PMe$_3$)$_2$Cp, Ru(PPh$_3$)$_2${HB(pz)$_3$}, RuCl(PCy$_3$)(η-MeC$_6$H$_4$Pr^i)

5.5 Cumulenylidenes

Extending the carbon chain of vinylidenes by one or three carbons provides linear cumulenated ligands, $L_nM=(C=C)_n=CR_2$ ($n = 1, 2$). These compounds have been studied in considerable detail, especially allenylidenes ($n = 1$) and arise in most cases from the rearrangement and dehydration (often spontaneous) of hydroxyalkyl alkynes or diynes (Figure 5.51). Various intermediates (Figure 4.18) may be envisaged (and occasionally isolated) which are related by elimination/addition of H^+ or hydroxide, *e.g.* γ-hydroxyalkynyl complexes have been isolated and subsequently dehydroxylated with Lewis or Brønsted acids.

This approach is inappropriate for the even-carbon ligands $L_nM=(C=$

C)$_n$=C=CR$_2$ and complexes of these ligands have yet to be isolated. However, this represents a synthetic hurdle rather than an indication that these complexes might not be stable. Indeed, butatrienylidenes, L$_n$M= C=C=C=CR$_2$, have been implicated in reactions of butadiyne (HC≡C– C≡CH) with transition metal complexes (Figure 5.52) and stable examples may be anticipated.

For the reactions of cumulenylidenes, a picture has begun to emerge wherein nucleophiles attack at C$_\alpha$, C$_\gamma$, C$_\epsilon$, etc., whilst electrophilic attack occurs at C$_\beta$, C$_\delta$, etc. (Figure 5.53). Thus, the reactions of vinylidenes with nucleophiles at C$_\alpha$ and electrophiles at C$_\beta$ fits within this scheme.

Figure 5.52 Higher metallacumulenes (cumulenylidenes); [Ru] = Ru(PPh$_3$)$_2$Cp; {Ru} = RuCl(dppm)$_2$

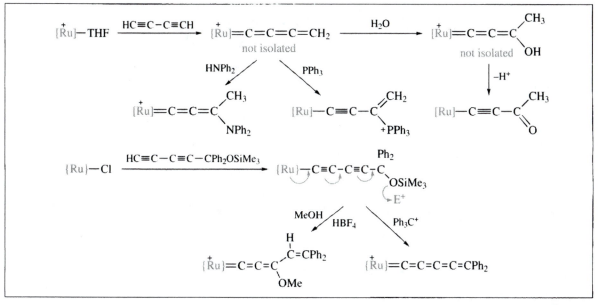

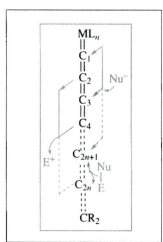

Figure 5.53 Regioselectivity of electrophilic and nucleophilic attack on metallacumulenes (n = 0, 1, 2, ...)

6

π-Coordination of C–C Multiple Bonds

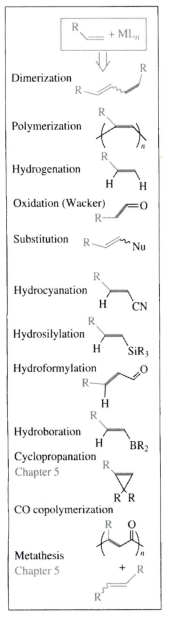

Dimerization

Polymerization

Hydrogenation

Oxidation (Wacker)

Substitution

Hydrocyanation

Hydrosilylation

Hydroformylation

Hydroboration

Cyclopropanation
Chapter 5

CO copolymerization

Metathesis
Chapter 5

Figure 6.1 Alkene complexes in organic (retro)synthesis ('n' isomers shown; iso or branched alternatives are also accessible)

Aims

By the end of this chapter you should have an appreciation of:

- The bonding of a variety of unsaturated organic molecules when π-coordinated to a transition metal (Dewar–Chatt–Duncanson model)
- The various synthetic routes to complexes of alkenes and alkynes
- The reactivity implications of π-coordination of unsaturated organic molecules to a transition metal
- How these may be applied to the design of productive catalytic cycles for alkene modification, including enantioselective examples

6.1   Introduction

Previous chapters concerned ligands which bind to a metal through one carbon atom ('C_1' ligands). Unsaturated organic substrates may also bind through two or more carbon atoms of a π-system. We will begin with simple 'C_2' examples, alkenes and alkynes, and then extend the principles to acyclic systems where the unsaturation extends through more carbon atoms. Carbocyclic polyenes will then be developed in Chapter 7. As with the ligands met previously, the coordination of C–C multiple bonds to a transition metal may greatly alter the reactivity of the molecule; however, this may often be harnessed by judicious choice of metal centres. Once appreciated, these principles may be exploited for the stoichiometric and catalytic functionalization of alkenes and alkynes. Figure 6.1 presents an impressive but by no means comprehensive array of selected alkene functionalizations that emerge in a retrosynthetic sense from alkene complexes, placing such complexes firmly within the organic chemist's synthetic armoury.

6.2   Alkenes

Only two years after Magnus reported the first coordination complex of a neutral molecule (NH_3), organotransition metal chemistry was born with the isolation by Zeise of *sal kalico-platinicus inflammabilis*, $K[PtCl_3(H_2C=CH_2)].H_2O$ (Figure 6.2). In the interim, all manner of unsaturated organic molecules have been shown to coordinate to transition metals through their unsaturated π-systems.

6.2.1   Bonding

The frontier orbitals of ethene are of suitable symmetry, energy and occupancy to interact synergically with the valence orbitals of a transition metal complex (Figure 1.18) as described by the Dewar–Chatt–Duncanson model. This considers the simple 2VE dative interaction (from a MO rather than a single donor atom), counter-balanced (to varying degrees) in pursuit of electroneutrality, by retrodonation into the alkene π* antibonding orbital. The relative extents of these two processes will depend on the energies of both the metal and (in contrast to CO) the alkene molecular orbitals. This, in part, determines the relative lability of the alkene coordination, although steric factors also play a part; increased alkene substitution typically leads to increased alkene lability. Alkene complexes are known for all of the transition metals in a wide range of oxidation states. By analogy with the π-acidity of CO, alkenes may even stabilize negative oxidation states, *e.g.* $[Fe(\eta\text{-}CH_2CH_2)_4]^{2-}$, in which case retrodonation from the metal assumes paramount importance.

6.2.2   Synthesis of Alkene Complexes

Generally, alkene complexes are prepared by simple ligand substitution processes beginning with the free alkene (Figure 6.3). Being π-acids, lower oxidation states are preferred and in some cases reactions with alkenes may result in reduction of the metal centre (sacrificial oxidation of excess alkene or solvent). A vacant coordination site is required, and the various means by which this coordinative unsaturation may be achieved have been met (Chapter 3). Notably, one class of labile precursors is alkene complexes themselves, when the electronic and steric properties of the introduced alkene relative to that displaced are taken into consideration. Loosely speaking, and with all else being equal, a more electronegative alkene will displace a more basic one, and a smaller

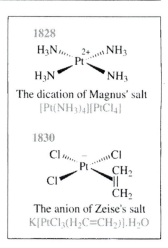

Figure 6.2 The births of coordination and organotransition metal chemistry

Alkene complexes are often useful as labile *in situ* sources (**synthons**) for coordinatively unsaturated metal complexes (Figure 3.14).

Signficant retrodonation (in addition to steric factors) may contribute to a barrier for alkene rotation about the metal–alkene axis (Table 1.4).

alkene will displace a larger or more heavily substituted one. Since equilibria may be involved, if the departing alkene is volatile, its removal from the system may direct the position of equilibrium in favour of the desired complex. This is illustrated by the wide use of the 2-methylpropene (isoprene) complex $[Fe(\eta\text{-}CH_2=CMe_2)(CO)_2Cp]^+$ {'Fp(η-$H_2C=CMe_2$)$^+$'} for the preparation of numerous alkene complexes of the Fp$^+$ fragment.

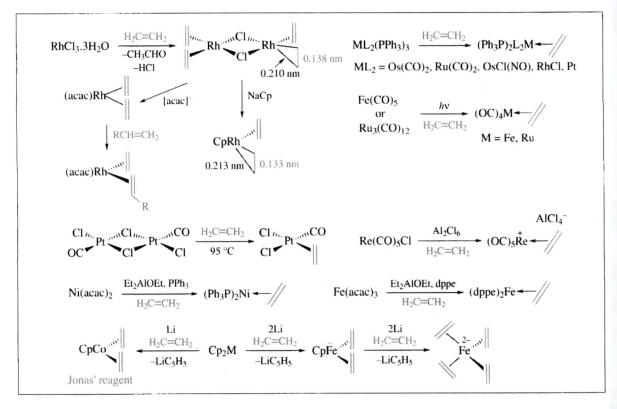

Figure 6.3 Selected syntheses of ethene complexes; acac = propane-2,4-dionate (acetylacetonate)

Alternatively, alkenes may also originate within a coordination sphere *via* modifications of existing ligands (*e.g.* β-M–H elimination, carbene rearrangement, Figures 2.31, 4.11, 4.25, 4.26, 4.28). Although already met, the most important of these are collected in Figure 6.4, because through microscopic reversibility, an appreciation of these processes provides retrosynthetic insights into the types of reaction that we might expect coordinated alkenes to undergo.

A key feature of alkene coordination is the loss of planar symmetry. Indeed, it is this loss of symmetry that allows the $v_{C=C}$ IR absorption of alkene complexes to be observed (IR inactive for the free molecule). Consistent with the synergic bonding description, coordination leads to a decrease in the C=C bond strength and accordingly a decrease in the value of $v_{C=C}$. The case of prochiral alkenes is of particular importance,

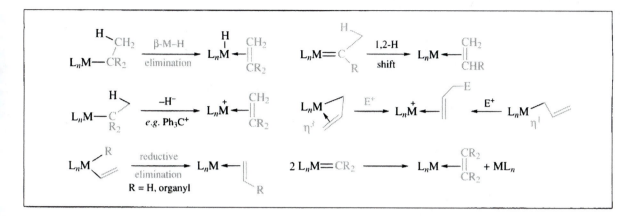

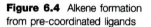

Figure 6.4 Alkene formation from pre-coordinated ligands

since either prochiral face of the alkene may coordinate to the metal centre (Figure 6.5). This is equivalent to the inclusion of a chiral centre within a cyclopropane ring. If the metal centre is itself chiral, then the energies (and kinetic implications) of the two possible facial coordinations will differ (diastereomeric reaction coordinates, Figure 6.5). This principle underpins the observed and technologically exploitable enantioselectivity of any organometallic process involving prochiral alkenes in combination with chiral transition metal centres (see below).

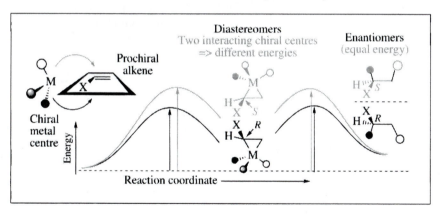

Figure 6.5 Diastereomeric reaction coordinates for the interaction of a prochiral alkene with a chiral metal centre

The chelate effect applies equally to alkenes as it does to classical ligands; dienes typically bind more effectively than monoalkenes. Perhaps the most widely employed of these are cycloocta-1,5-diene (cod) and norbornadiene (nbd) (Figure 6.6). Conjugated 1,3-dienes tend to bind more efficiently than non-conjugated α,ω-dienes (see below). Transition metal centres are capable of isomerizing alkenes (see below, Figure 6.7), and accordingly non-conjugated dienes may be isomerized to their conjugated 1,3-isomers for more favourable coordination. Conjugated di- and polyenes will be returned to later in the chapter.

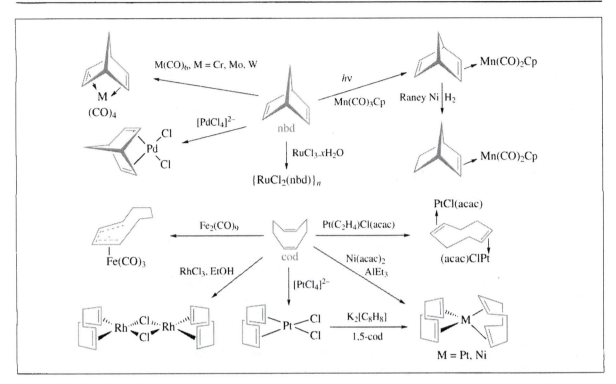

Figure 6.6 Complexes of cyclic dienes; cod = cycloocta-1,5-diene; nbd = norbornadiene (bicyclo[2.2.1]hepta-2,5-diene)

Figure 6.7 (Below) Metal-mediated alkene isomerization *via* (a) allyl or (b) hydride intermediates

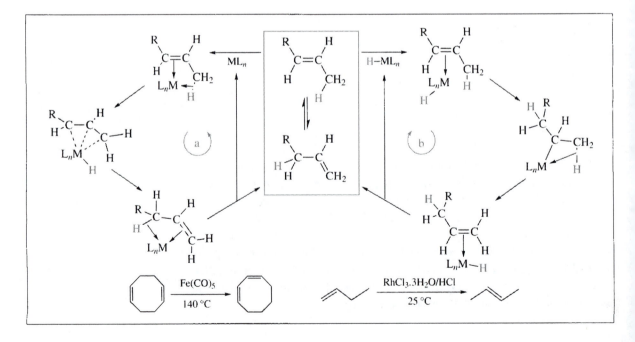

6.2.3 Reactions of Coordinated Alkenes

Alkene Rearrangements

The metal-mediated rearrangement of alkenes is an immensely broad topic in its own right; only the more general principles will be discussed. The various alkene rearrangement possibilities all rely on C–H activation processes, which in turn necessitate that the C–H bond approaches the metal centre. The most important processes therefore involve hydrogen atoms bound either directly ('vinylic') or adjacent ('allylic') to the alkene. For coordinatively saturated alkene complexes, formation of the simple π-complex ensures that the alkene is within the coordination sphere; however, rearrangement to an intermediate in which the vinylic C–H σ-bond coordinates prior to oxidative addition allows the metal centre not to exceed 18VE (Figure 6.7). For allylic C–H activation, a vacant coordination site *cis* to the alkene will facilitate a concerted mechanism, and evidence in support of this comes from the isolation of alkene complexes wherein the allylic C–H bond coordinates agostically (Figure 2.24). Once an allyl ligand (Chapter 7) is generated, reductive elimination of the hydride ligand may occur at either end of the allyl ligand to generate the original alkene or, alternatively, a new isomer. Amongst the most effective metal complexes for the isomerization of alkenes are metal hydrides, which, in addition to the previous mechanism, may insert the alkene to provide σ-alkyl ligands. These may then isomerize *via* a series of β-M–H elimination/reinsertion steps. The equilibria in such sequences are generally dictated by steric factors; alkene coordination is less favoured for polysubstituted alkenes; metal bond strengths decrease with α-branching (Table 1.5, Figure 6.7).

In the previous chapter it was noted that alkene metathesis may be catalysed by some pre-catalysts which have neither alkylidene nor metallacyclobutane ligands. In such systems, the alkene itself serves as the alkylidene source, presumably through rearrangement to a carbene *via* a sequence of hydrogen shifts. Although this is unlikely to be directly observed in the most active catalysts, examples of this rearrangement have been observed in stoichiometric reactions. Generally, however, the reverse alkylidene → alkene rearrangement is more common.

Nucleophilic and Electrophilic Attack

It has already been seen that the reactivity of alkylidene and alkylidyne complexes may be dictated by the nature of both the metal and the carbon substituents. These principles apply equally to alkene coordination (Figure 6.8). Two metal–ligand fragments serve to illustrate these principles in the activation of alkenes to electrophilic and nucleophilic

Figure 6.8 (a) Electrophilic (E^+) and (b) nucleophilic (Nu^-) attack at alkene complexes

reagents: 'Ru(dmpe)$_2$' (Figure 6.9; electron-rich, neutral, zerovalent, σ-donor co-ligands) and 'Fe(CO)$_2$Cp$^+$'(Figure 6.10; electron-poor, cationic, divalent, competitive π-acid co-ligands).

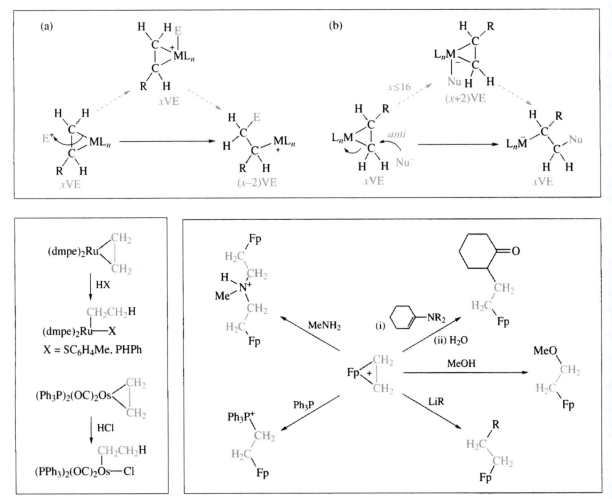

Figure 6.9 Electrophilic attack at coordinated ethene

Figure 6.10 (Right) Nucleophilic attack at coordinated ethene; Fp = Fe(CO)$_2$Cp

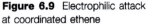
Many nucleophiles (e.g. phosphines) are not strong enough to react with simple free alkenes, but readily react with alkenes coordinated to suitably electrophilic metal centres.

Nucleophilic attack at coordinated alkenes might be viewed as analogous to the organic chemistry of bromonium ions ('alkene complexes' of Br$^+$), and is favoured by positively charged metals and/or those bearing strongly π-acidic co-ligands. The range of nucleophiles is enormous (Figure 6.10), with the caveat that many nucleophiles can also serve as potential ligands. This may be a problem if the alkene coordination is labile.

The product of nucleophilic attack at a coordinated alkene is generally an alkyl ligand that bears the nucleophile β (and *anti*) to the metal centre. Since alkyls with protons β to the metal centre are prone to β-M–H elimination (releasing an alkene), this may be built into a catalytic cycle for the nucleophilic functionalization of alkenes. One application of this is the catalytic oxidation of ethene to ethanal (acetaldehyde) by [PdCl$_4$]$^{2-}$ (Wacker process; Figure 6.11 for the reaction in D$_2$O). In this process, divalent palladium is reduced and must be re-oxidized by a combination of air (sacrificial oxidant, *i.e.* consumed) and copper(II) chloride (catalytic, *i.e.* regenerated oxidant). If methanol is used in place of water as the nucleophile, methyl vinyl ether is obtained stoichiometrically (Figure 6.12).

Figure 6.11 The Wacker process for ethene oxidation

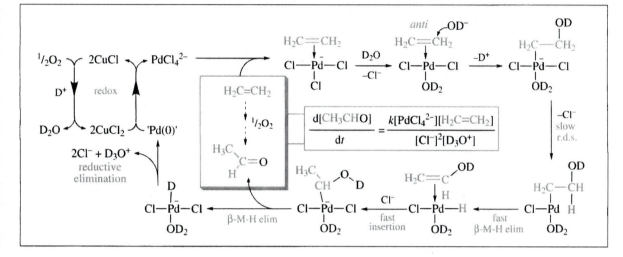

In some instances (<18VE) it is not always certain that nucleophilic attack occurs directly at the coordinated alkene, since attack at the metal centre followed by migration of the nucleophile to the alkene (alkene insertion, see below) may result in the same overall reaction product (Figure 6.8).

Electrophilic attack will be favoured by metals in low oxidation states, with few or no competitive π-acceptor ligands. In such situations, retrodonation from the metal loads the alkene with excessive electron density which is attractive to an external electrophile. As in the case of nucleophilic attack, it may not always be clear, however, whether electrophilic attack has occurred directly at the alkene, or alternatively at the metal centre followed by insertion of the alkene into the metal–electrophile bond (Figure 6.8). This is a real possibility, since the factors that activate the alkene to electrophilic attack are also factors that render the metal centre potentially nucleophilic. Furthermore, the electrophilic conversion of an alkene to a β-functionalized alkyl deprives the metal centre of 2VE; this may make it prone to β-M–E(H) elimination (unless blocked by coordination of the conjugate nucleophile), obscuring the mechanistic detail (Figure 2.25).

The microscopic reverse of the β-metal hydride (β-M–H) elimination

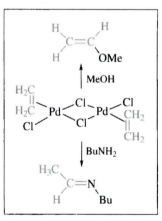

Figure 6.12 Stoichiometric reactions of Pd$_2$Cl$_4$(η-C$_2$H$_4$)$_2$

The insertion reaction is not limited to metal hydrides; a variety of reactive metal–element (e.g. M–C, M–Si, M–B) bonds will insert cis-coordinated alkenes.

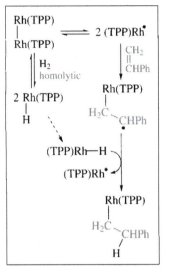

Figure 6.13 Radical route for alkene hydrometallation

reaction is alkene insertion, a special case of the more general insertion of unsaturated ligands into metal–hydride bonds (see Chapter 2, Figures 2.30, 2.31). By far the majority of metal–hydride alkene insertion processes occur *via* concerted four-membered transition states (Figure 2.31); however, it should be noted that alternative mechanisms exist which may lead to the same overall stoichiometric result. Of these, radical insertion processes are of note (*e.g.* Figure 6.13).

The insertion of alkenes into metal–hydride bonds is of central importance to many alkene functionalization processes, many of which are catalytic in nature. Depending on the co-ligands present (or subsequently introduced), reductive elimination of the resulting σ-organyl ligand may be possible. This will be illustrated for the case of alkene hydrogenation and then developed to include many of the variants illustrated in Figure 6.1. The hydrogenation of alkenes catalysed by Wilkinson's catalyst, RhCl(PPh$_3$)$_3$, is depicted in Figure 6.14. This is a mechanistic simplification; many subtle equilibria exist involving chloride and phosphine dissociation. Furthermore, an alternative sequence involving alkene coordination *prior* to hydrogen oxidative addition would also lead to the same crucial intermediate with hydride *cis* to the alkene. The relative rates of these two routes will depend on the particular (and numerous!) variants of Wilkinson's original rhodium(I) catalyst.

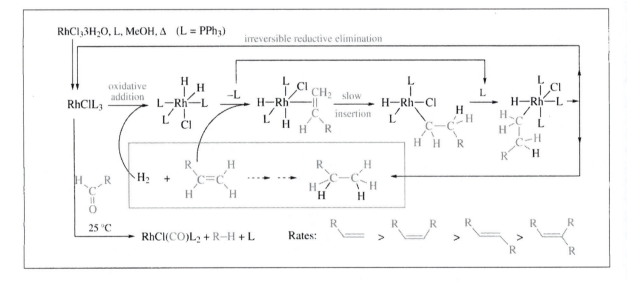

Figure 6.14 Alkene hydrogenation by RhCl(PPh$_3$)$_3$ (Wilkinson's catalyst)

Important features of RhCl(PPh$_3$)$_3$ include (i) sterically induced lability of one phosphine; (ii) solvent-dependent lability of the chloride ligand; (iii) facile interconversion between two electronically advantageous oxidation states and geometries (d^8 ML$_4$ square planar and d^6 ML$_6$ octahedral). The lability of chloride or phosphine ligands allows for the generation of vacant coordination sites for alkene coordination and oxidative addition processes. The initial step involves loss of phosphine and dihydrogen oxidative addition to form the coordinatively unsaturated hydride complex. This coordinates the alkene followed by a rate-determining insertion of the alkene into one Rh–H bond followed by an irreversible reductive elimination of an alkane to regenerate the catalyst. Wilkinson's catalyst was remarkable at the time of its discovery (1965); perhaps more remarkable is that, despite the intervening 35 years of intense research into alternatives, it still enjoys wide use. The key features that make it attractive are therefore noteworthy. Firstly, the synthesis from commercial $RhCl_3.3H_2O$ and PPh_3 (methanol reflux) is high yielding and simple. Low concentrations (<1 mmol dm^{-3}) of the complex will effectively hydrogenate alkenes at ambient temperatures under low pressures (1 bar, 10^5 Pa) of hydrogen. Given the importance of steric factors in the organometallic chemistry of both alkenes and σ-alkyls, it is not surprising that Wilkinson's catalyst shows the indicated rate dependence on the degree of alkene substitution (these observations generally apply to many other hydrogenation catalysts). It is also tolerant of a wide range of functional groups (esters, ketones, nitro compounds and carboxylic acids). Aldehydes are, however, decarbonylated to the corresponding hydrocarbon in a stoichiometric (and therefore prohibitively expensive) process under mild conditions. At higher temperatures (>200 °C) the reaction does become catalytic; however, by employing cationic complexes, the lability of the resulting coordinated CO may be increased, favouring dissociation and regeneration of the catalyst. Thus the complex $[Rh(triphos)(CO)]^+$ effects catalytic decarbonylation at 100 °C (Figure 6.15).

Figure 6.16 shows some of the commonly encountered (pre)catalyst

Figure 6.15 Catalytic aldehyde decarbonylation

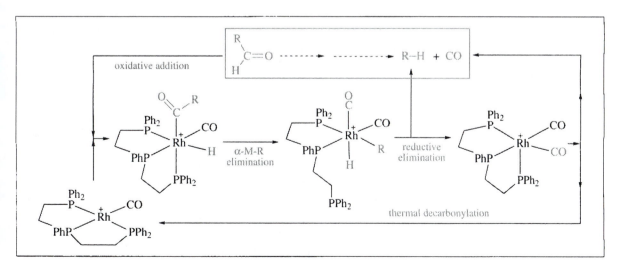

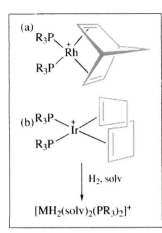

H_2, solv

$[MH_2(solv)_2(PR_3)_2]^+$

Figure 6.16 M^I (M = Rh, Ir) alkene hydrogenation pre-catalysts

variants which employ the basic rhodium or iridium (I/III) reaction manifold. The cationic complexes (a) or (b) react with hydrogen in donor solvents (THF, acetone, ethanol, acetonitrile) *via* hydrogenation of the cod or nbd ligand to generate $[MH_2(solvent)_2(PR_3)_2]^+$ (M = Rh, Ir) as the active catalyst. A wide variety of derivatives are available by variation of the phosphine (PR_3) and solvent. These (halide-free) catalysts offer the advantage that, being cationic, any potential donor group which the alkene may possess (amides, alcohols, esters, *etc.*) may also bind to the metal. This becomes particularly important for enantioselective alkene hydrogenation (see below).

Two important alternatives to dihydrogen oxidative addition also exist and may operate in some systems. Firstly, heterolytic cleavage of a dihydrogen ligand (H^+, H^-) by base is implicated in the formation of the hydrogenation catalyst $RuHCl(PPh_3)_3$ from $RuCl_2(PPh_3)_3$, hydrogen and Et_3N. Alternatively, some complexes, *e.g.* $[Co(CN)_5]^{3-}$ and $[Rh(TPP)]_2$, may homolytically cleave dihydrogen (H, H) to provide $[CoH(CN)_5]^{3-}$ (Figure 6.17) and RhH(TPP) (Figure 6.13), respectively.

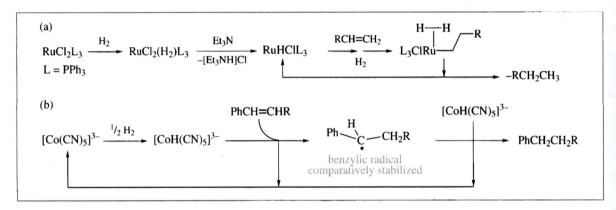

Figure 6.17 (a) Heterolytic and (b) homolytic activation of hydrogen

Hydrogenation is of enormous synthetic potential, but represents only one dimension of Figure 6.1. Many other functional groups may be introduced by essentially the same process. Each of these requires the formation of the key alkyl intermediate, but varies in the ways in which this is ultimately cleaved from the metal. Alkene hydrocyanation, hydrosilylation and hydroboration represent key routes to important classes of organic commodity chemicals, which may be effectively catalysed by transition metal complexes. These processes differ essentially only in the nature of the group that is ultimately eliminated with the alkyl formed by alkene insertion. In place of hydrogen, HCN, a silane or a borane is used for the oxidative addition step. Figure 6.18 provides a generalized mechanistic model for these three processes (only the route to the *n* isomer is shown). The danger of generalization, however, is that the actual sequence and rate-determining steps may differ, depending on the particular catalyst and substrate, and whether insertion of the alkene occurs into the M–H or M–X bond. Whatever their actual order, the basic steps remain fundamental.

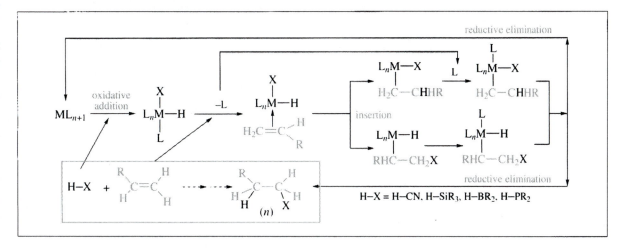

Figure 6.18 Generalized catalytic addition of 'H–X' to alkenes

Alkene hydrogenation may be diverted, if the reaction is carried out in the presence of carbon monoxide, leading to the hydroformylation process (Figure 6.19). The intermediate alkyl complex, being coordinatively unsaturated, may add CO to form a *cis* alkyl-carbonyl complex in equilibrium (Figure 3.24) with the corresponding acyl complex. Reductive elimination of aldehyde finally regenerates the catalyst. Numerous metal hydride complexes (or suitable precursors) have been shown to mediate this reaction. Two examples serve to illustrate key points, *viz*. $RhH(CO)(PPh_3)_3$ and $Co_2(CO)_8/H_2$ [in equilibrium with $CoH(CO)_4$; Figure 3.11]. Firstly, in contrast to alkene hydrogenation (H, H), for hydroformylation the addenda (H, CHO) are different and accordingly isomeric products (*n*- *vs.* iso) may arise. At present, commercial interests generally favour the formation of the *n*-isomer, *i.e.* where the formyl group adds to the terminal carbon (C^1) of the alkene, rather than the iso isomer (formyl resides on C^2). The more sterically congested coordination sphere of the rhodium catalyst favours formation of the *n*-alkyl isomer, whilst the less sterically cumbersome cobalt complex can accommodate both *n*- and branched (iso) alkyl ligands, leading to a loss of selectivity. Furthermore, high pressures of CO (and associated plant costs) are required to prolong the life of the cobalt catalyst. Thus for rhodium, catalyst cost is offset by selectivity. It should be noted that alkene hydrogenation is thermodynamically more favourable than hydroformylation (*e.g.* by 34 kJ mol⁻¹ for ethene) and so a requirement of the catalyst is that it kinetically diverts the reaction away from simple hydrogenation.

The *n*/iso ratio for the cobalt system can be significantly increased by

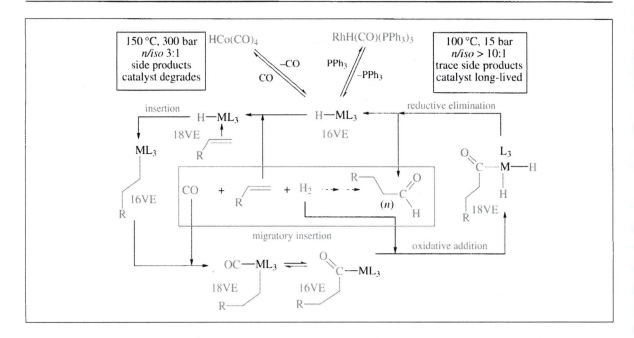

Figure 6.19 Alkene
hydroformylation

The rhodium process has been in
commercial use since 1976. In
the interim the increasing
importance of environmental,
health and safety considerations
in industrial processes make the
use of less-extreme reaction
conditions, catalyst longevity and
the minimization of side-product
formation of crucial importance.

inclusion of bulky ligands, *e.g.* 9:1 for the complex $HCo(CO)_3(PBu_3)$. Since the simple $HCo(CO)_4$ catalyst degrades *via* CO loss (depositing Co metal), introduction of stabilizing ligands may help to increase catalyst lifetime but compromise the catalyst activity. Further current variants include polymer-bound or water-soluble versions of the rhodium system (see Figure 2.6).

Enantioselective Alkene Reduction

So far, we have only considered 'generalized' alkenes and not addressed the questions of stereochemistry that arise when prochiral alkenes are functionalized with loss of planar symmetry. Hydrogenation (and hydrocyanation, hydroformylation, hydrosilylation and hydroboration) of a prochiral alkene must lead to a racemic mixture of products if the metal catalyst and reaction medium are achiral. Coordination of a prochiral alkene to a chiral metal centre, however, leads to diastereomeric complexes (different energies). Accordingly, the rates of both the formation and any subsequent reactions will be different (Figure 6.5). Ideally, it would be useful if the rates for the two diastereomeric reaction coordinates were sufficiently different that productive enantioselectivity was observed; indeed, this is often the case. Enantioselectivity may arise from the initial (ideally reversible) coordination stage, or during *any* of the subsequent reactions prior to or including the rate-determining step. For this reason, the entire reaction sequence must be considered, and the accumulation of one particular diastereomeric intermediate in sufficient quantities for spectroscopic observation *is not* necessarily indicative of the predominant reaction. Indeed, the absence (or minority) of the alternative diastereomer may simply reflect its greater rate in a subsequent (selectivity determining) step, which drains the equilibrium manifold in the direction of the final predominant enantiomer. It is a guiding principle for the study of catalysis that however much we might hope to observe (characterize and understand) catalytic intermediates, the more effective the catalysis, the less likely that the *key* intermediates will persist long enough to be observed.

Perhaps the greatest effort has been directed towards the design of chiral metal catalysts capable of enantioselective hydrogenation of prochiral alkenes. Fortunately, knowledge gained in such studies (and the resulting catalysts) may often be applied to the broader scheme of enantioselective alkene functionalization in general, given the generality suggested by Figure 6.18. Figure 2.3 presented one approach to enantio-selective catalysis involving the use of chiral phosphine co-ligands (see also Figure 2.10). The Monsanto company applied this to the development of a chiral catalyst for the enantioselective hydrogenation of the prochiral dehydroalanine precursor to L-dopa, used in the treatment of Parkinson's disease (Figure 6.20). The active catalyst is generated from the cod complex (pre-catalyst) by *in situ* hydrogenation of the diene to (non-coordinating) cyclooctane in a pre-activation step (see Figure 6.16). This liberates two potential coordination sites which are occupied by labile solvent in the resting state of the catalyst. This particular case is slightly more complex than simple alkenes in that further coordination of the amide group to the cationic centre enhances the enantioselectivity. Enantioselective catalysis is routine in biological systems; a feature of this synthetic approach is that by inverting the chirality of the phosphine, the alternative enantiomer of the product may be selectively prepared, if required.

Figure 6.20 Monsanto L-dopa synthesis: Co-ligand dependence

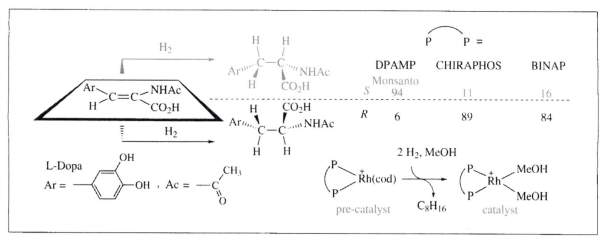

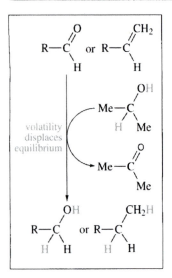

Figure 6.21 Transfer hydrogenation of alkenes and aldehydes

The principles of asymmetric alkene hydrogenation may also be extended to other C=X double bonds, the greatest success being achieved to date with aldehyde, and to a lesser extent imine, hydrogenation. The steps in such a reductive sequence parallel those for alkenes with the exception that an alkoxide, rather than an alkyl, complex results from insertion of the aldehyde. This is the reverse of the β-M–H elimination reaction observed for alkoxides (Figure 4.26), and which underpins the numerous syntheses of metal hydride complexes in basic alcoholic solvents. A clever variant of this equilibrium process, known as transfer hydrogenation, makes use of the different volatilities of propan-2-ol and propanone (acetone) and the equilibrium nature of the reaction manifold. Thus dehydrogenation of propan-2-ol provides both volatile propanone and an intermediate dihydride (or dihydrogen) complex for use in productive hydrogenation of the desired aldehyde. Aldehydes and alkenes are generally more prone to hydrogenation than ketones; hence the reduction of these substrates may be carried out without the need to use gaseous hydrogen as a substrate (Figure 6.21).

Alkene Dimerization and Oligomerization

If metal complexes (in particular hydrido or alkyl complexes) are able to coordinate more than one alkene, the possibility of alkene coupling arises. Amongst the various mechanisms that can account for this coupling, the two shown in Figure 6.22 each involve β-M–H elimination as the key step, which ultimately leads to (possibly catalytic) alkene dimerization. This elimination naturally requires a vacant coordination site, and if this is available to further alkene coordination, oligomerization or polymerization may occur. This is highlighted by the phosphine co-ligand dependence of the nickel example in Figure 6.22, where the use of PBu^t_3 leads exclusively to oligomerization. The intermediacy of metallacyclopentanes is suggested by the reaction of $Fe(CO)_5$ with tetrafluoroethene (Figure 6.22), which provides a metallacycle which is stable, primarily due to a combination of the Fe–C bond strength enhancement offered by α-fluoro substituents (Table 1.5) and the relative lability of Fe^0–F (HSAB) *vs.* C–F bonds which will discourage β-M–F elimination. Indeed, perfluoroalkenes are particularly prone to insertion reactions with metal–ligand bonds.

Aluminium alkyls will polymerize ethene to polyethene, albeit inefficiently. During studies of this process, Ziegler found that a nickel contaminant resulted in the exclusive formation of butene, *i.e.* dimerization (see above). This prompted a study of the reactions of various transition metal complexes in combination with aluminium alkyls. Recall (Figure 4.6) that aluminium alkyls readily transfer their σ-organyl groups to the more electronegative transition metals. Amongst these, titanium

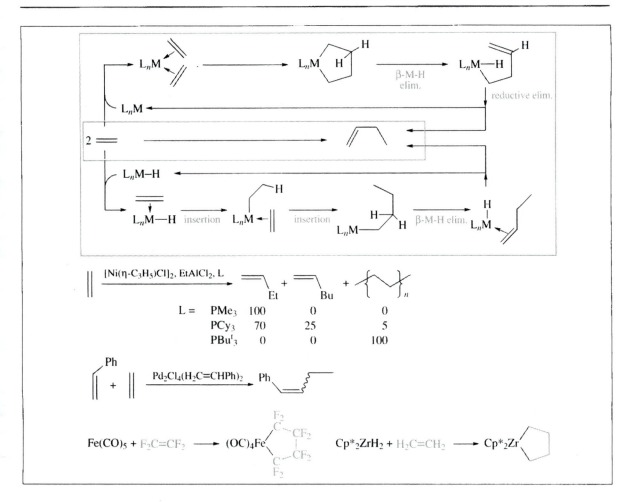

Figure 6.22 Catalytic alkene dimerization and models for intermediates

and zirconium halides provided crystalline, high-density polyethene under mild conditions (50 °C, 10 bar). These desirable properties result from the low degree of alkyl branching along the polymer chain, which is otherwise commonly encountered in non-transition metal polymerization processes (thermal, radical, anionic, cationic). Natta extended this process to propene with either TiCl₃/Et₃Al or VCl₄/Et₂AlCl and found that different types of polypropene result, differing in the relative orientation of the polymer spine methyl substituents (Figure 6.23). For free radical polymerization the relative orientations of substitutents are random (atactic). For the titanium system, all substituents lie on the same side (isotactic) of the hydrocarbon chain (Fischer projection sense). For the vanadium system, however, the substituents lie on alternating sides (syndiotactic). Since the different forms atactic, isotactic and syndiotactic have different mechanical and physical properties, the control of tacticity is crucial for subsequent applications (Figure 6.23).

For syndiotactic polypropene, one currently employed variant on the

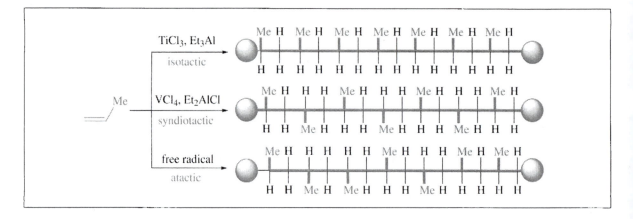

Figure 6.23 Control of
polypropene tacticity

Ziegler–Natta system involves a combination of α-TiCl$_3$, TiCl$_4$ and ethyl benzoate doped (*ca.* 1%) in bulk MgCl$_2$, which is then activated with Et$_3$Al. The activity of this somewhat ill-defined mixture is sufficiently high that it is unnecessary to remove the traces of catalyst from the bulk commodity polymer. Alternatives include the Reduced Phillips and Union Carbide catalysts, which are based on the incorporation of low-valent chromium species into silica gel, wherein the alkene itself *somehow* is converted into the propagating alkyl, without requiring organoaluminium additives. Whilst these mixtures are justified on the grounds of activity and economy, they are frustrating on an academic level. Ignorance of the true nature of the active site(s) requires that any further refinement be based on empirical ('trial and error') iteration. Molecular alkene polymerization catalysts have been developed, however, and are finding increasing industrial application, whilst providing insight into the mechanism of polymerization. These all have in common a situation where a coordinatively unsaturated alkyl complex is generated, the most effective involving 14VE (and therefore electrophilic) metal centres. Many transition metal halides react with 'methylaluminoxane' (MAO) to provide active Ziegler–Natta-like catalysts. MAO is a poorly defined partial hydrolysis product of Al$_2$Me$_6$, which nevertheless effectively serves two purposes: (i) as an alkyl transfer reagent to the transition metal; and (ii) as a Lewis acid which abstracts one alkyl to generate a vacant coordination site. Alternatively, dialkyls may be isolated using more conventional alkylating agents and then treated with an alkyl abstracting agent [electrophilic cleavage, *e.g.* B(C$_6$F$_5$)$_3$, Ph$_3$C$^+$, or a weak proton donor such as PhNMe$_2$H$^+$, Figure 4.28]. When the resulting alkyl complex is cationic, it is essential for high activity that the counter anion does not compete effectively with the alkene for the requisite coordination site (see Figure 2.27b), whilst the inclusion of sterically demanding co-ligands ensures that the catalyst does not deactivate *via* bimolecular decomposition routes (not a problem for heterogeneous catalysts). Since the relative stereochemistries of the carbon centres of the polymer spine dictate the tacticity, the inclusion of chirality at the metal centre may assist in controlling the enantioselectivity of C–C bond formation (Figure 6.5). Figure 6.24 illustrates (a) some representative molecular Ziegler–Natta-type catalysts and (b) metal halide complexes which serve the same purpose in combination with MAO.

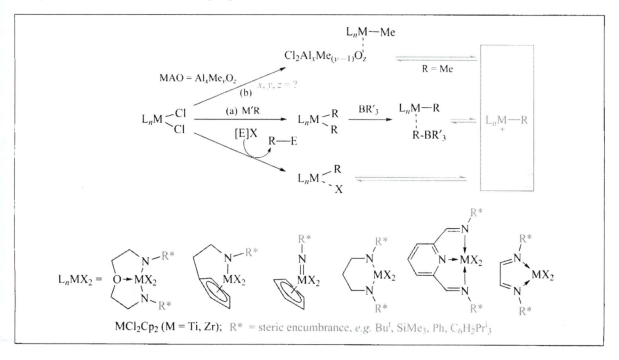

MCl$_2$Cp$_2$ (M = Ti, Zr);   R* = steric encumbrance, e.g. Bu^t, SiMe$_3$, Ph, C$_6$H$_2$Pr$^i_3$

Although others have been proposed, the most widely accepted mechanism for Ziegler–Natta polymerization was proposed by Cossee (Figure 6.25) and might be viewed as tangental to Figure 6.22. If the initially formed and coordinatively unsaturated alkyl complex may co-ordinate a further alkene rather than undergoing β-metal hydride elimination, then this will lead to polymer growth. Thus if alkene co-ordination is faster than β-elimination, productive polymerization occurs. If, however, β-M–H elimination competes, then termination occurs to generate a metal hydride. This does not *necessarily* destroy the catalyst(s), since it may (depending on the particular catalyst) in principle insert a further equivalent of the monomer to regenerate a new alkyl complex, and the polymerization may reinitiate to form a new propagating chain.

Figure 6.24 (a) Well-defined molecular alkene polymerization initiators; (b) MAO-activated initiation [E = CPh$_3$, H, HNR$_3$; X = weakly coordinating anion (see Figure 2.27b); R' = Ph, C$_6$F$_5$, C$_6$H$_3$(CF$_3$)$_2$]

6.3   Di- and Polyenes

It was noted above that conjugated 1,3-dienes generally coordinate more effectively to transition metals than do α,ω-dienes. This may be traced

Figure 6.25 Cossee mechanism for alkene polymerization

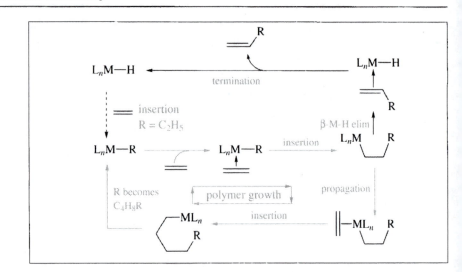

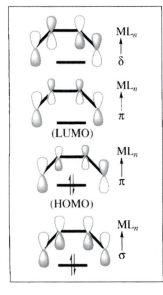

Figure 6.26 Butadiene frontier molecular orbitals

to a better energetic balance of frontier orbitals for the conjugated system with those of a transition metal. Figure 6.26 illustrates the molecular π-orbitals of butadiene in addition to the symmetries (σ, π, δ) these present to a metal as it approaches one face (Figure 6.26). By far the most common coordination mode for butadienes involves the η^4-coordination of the *s-cis* form. Notably, free butadienes typically adopt the energetically more favourable (12 kJ mol^{-1} for butadiene) *s-trans* form. For earlier transition metals, however, coordination in the s-*trans* form is prevalent (Figure 6.27). This is especially true in the case of bent metallocenes (Figure 6.28, see Chapter 7), where the frontier orbitals of the metallocene are essentially restricted within one plane (between the two Cp rings) and are therefore better oriented for overlap with the s-*trans* form. An alternative coordination of the s-*cis* form involves the formation of two σ-M–C bonds to the termini of the butadiene and the localization of a C–C double bond between the two internal carbon atoms. This double bond may coordinate to the metal centre (σ,σ',π), it may remain remote (σ,σ') or, as is occasionally observed, there may be a weak interaction. Thus a continuum of coordination modes is possible, depending on the requirements of the metal centre. This even extends to *ortho*-xylylene ligands, where the aromaticity of the C_6H_4 unit might be expected to be compromised by coordination. In solution there is often evidence to suggest that a metal may migrate between the faces of η^4-coordinated butadienes *via* the intermediacy of such σ,σ'-isomers (Figure 6.27).

Some metals show such a pronounced tendency to adopt η^4-butadiene coordination that this may override the driving force to aromaticity. Selected complexes of d^8 M(CO)$_3$ (M = Fe, Ru) fragments illustrating this point are included in Figure 6.29. Phenol is far more stable than its keto tautomer (cyclohexadienone), which may be stabilized, however, by coordination to iron(0). In a similar manner, divinylbenzene coordinates two 'Fe(CO)$_3$' fragments, disrupting the aromaticity. The complex [Fe(η-C$_6$Me$_6$)$_2$]$^{2+}$ (see Chapter 7) on reduction to iron(0) results in loss of the aromaticity of one arene to provide a *tetrahapto* butadiene coordination. This type of η^4-coordination is often encountered for complexes of naphthalene ($4n + 2 = 10\pi$), where some aromaticity ($4n + 2 = 6\pi$) is retained in the second ring of the fused arene. The drive towards buta-1,3-diene coordination may be sufficient to favour rearrangement of non-conjugated dienes, or the ring opening of cyclopropylalkenes (Figures 6.2, 6.6, 6.29).

We have previously emphasized the activation of organic molecule *via* coordination. The converse may also be true, such that transition metals may be used as protecting groups for reactive unsaturated functional groups. Thus η^4-metal coordination of a butadiene system deactivates it towards Diels–Alder reactions. Friedel–Crafts acylation of free polyenes

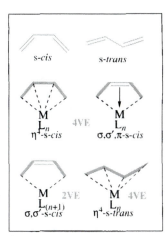

Figure 6.27 Butadiene coordination modes

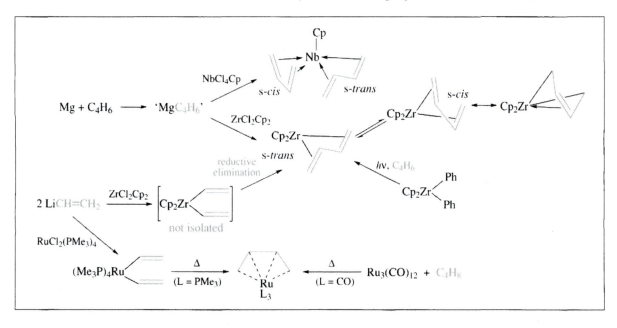

Figure 6.28 Synthesis of butadiene complexes

is a somewhat indiscriminate process; however, this proceeds cleanly when the diene is coordinated to the Fe(CO)$_3$ unit. Iron tricarbonyl adducts of butadienes are comparatively stable towards aerial oxidation, allowing ease of handling; decomplexation may be achieved, however, by stronger oxidants (*e.g.* Ce^{IV}) or reagents which abstract CO, leading to decomposition of the complex, *e.g.* Me$_3$N–O, PhI–O, RC≡N–O (see Figure 3.15). The special case of cyclobutadienes (anti-aromatic in the free state) will be discussed in the next chapter.

1,3,5-Trienes may also coordinate to a metal centre, providing a total of 6VE to the electron count; however, this is only favourable when the

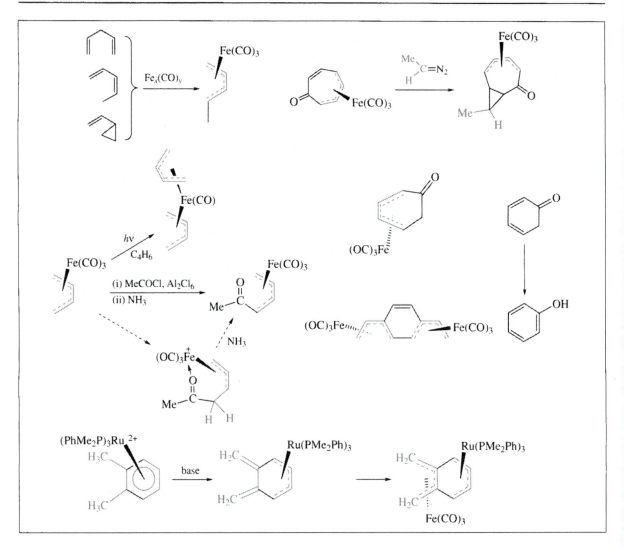

Figure 6.29 Butadiene coordination chemistry

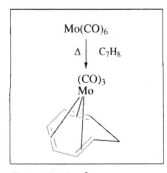

Figure 6.30 η^6-Cycloheptatriene coordination

spatial arrangement of the three double bonds allows all three to assume facial stereochemistry, *i.e.* if the triene has the *cis-cis* arrangement, *e.g.* in the case of cycloheptatrienes (Figure 6.30).

Cyclooctatetraenes are an intriguing class of organometallic ligand that could equally well be discussed in this or the following chapter. Free cyclooctatetraene is a non-planar (tub-shaped) cyclic polyene; a planar structure would constitute an anti-aromatic ($8\pi \neq 4n + 2$) system. The molecule may be reduced (2VE), however, by Group 1 metals to provide the aromatic (10π) dianion. For metals of Groups 6–10, the polyene behaves as a non-planar ligand, coordinating through one, two or three of the double bonds (Figure 6.31), depending on the electronic requirements of the metal centre(s). Since this leaves one or more double bonds pendant, the possibility of bridging two metals arises. Earlier transition metals with lower d occupancies (or actinides which fall outside the 18VE regime) offer the possibility of coordinating C_8H_8 in a planar manner since the metal is able to accept the full 8VE potentially on offer from a C_8H_8 ligand (Chapter 7).

6.4   Allenes

The π-system of allene ($H_2C=C=CH_2$) is not conjugated at the central carbon owing to the orthogonality of the two central carbon p-orbitals. Thus it is not favourable for the allene to provide 4VE to one metal without severe distortion. Many complexes of allenes occur, however, in which one double bond coordinates to a metal centre. In principle, the exocyclic double bond remains available for coordination to a second metal centre, resulting in a dimetallaspiropentane structure (Figure 6.32). If we consider the metallacyclopropane resonance form, then it appears that there are two types of metal–carbon σ-bond: an alkyl and a vinyl. These may therefore be expected to show different (regioselective) reactivities. Thus, as with simple alkenes, coordination to a metal centre may result in activation to either nucleophilic or electrophilic attack and in both cases the external reagent ultimately resides on the terminal carbon to provide an alkenyl ligand rather than attack at the central carbon to give an allyl ligand (see Chapter 7). Insertion of allenes into metal–hydride bonds, however, typically results in the formation of allyl complexes, although it should be noted that the allyl and alkenyl isomers may be related by β-M–H elimination–alkene re-insertion processes (Figure 6.32).

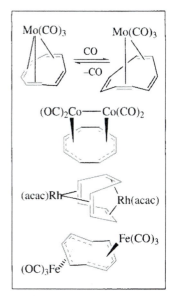

Figure 6.31 Non-planar cyclooctatetraene coordination

Figure 6.32 Allene coordination chemistry

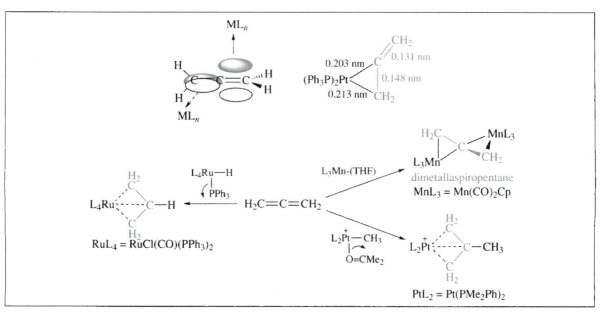

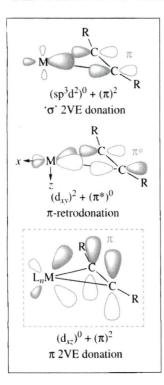

$(sp^3d^2)^0 + (\pi)^2$
'σ' 2VE donation

$(d_{xy})^2 + (\pi^*)^0$
π-retrodonation

$(d_{xz})^0 + (\pi)^2$
π 2VE donation

Figure 6.33 Metal–alkyne orbital interactions

6.5   Alkynes

Much of the coordination chemistry of alkynes may be viewed as analogous to that of alkenes within the context of the Dewar–Chatt–Duncanson model (Figure 6.33). Similar arguments apply to the factors which dictate the relative importance of σ-donation and π-retrodonation, and the effect of these on the reactivity of the coordinated alkyne. A further dimension is included, however, by the presence of the second set of π-orbitals which are orthogonal to those used in synergic bonding to the metal. Two important situations arise where these orbitals may become involved. Firstly, a second metal centre may coordinate to the alkyne by making use of these, in which case a dimetallatetrahedrane structure is adopted. This may be supported by a direct metal–metal bond (Figure 6.34); however, this is not always essential. In this coordination mode the alkyne provides a total of 4VE to the overall electron count, with the C–C bond perpendicular (transverse) to the M–M bond. If, however, the alkyne coordinates parallel to the M–M bond, then it only provides 2VE and might be viewed as a dimetalla-alkene. Interconversion between these two bonding modes is occasionally observed. A third possibility arises if the alkyne is bound to a trinuclear cluster that might be considered as a hybrid of these two, in which the available 4VE are shared between three metals. Remarkably, this may lead to eventual cleavage of the remaining C–C bond and formation of a cluster with two bridging alkylidyne ligands, which together provide a total of 6VE to the overall count.

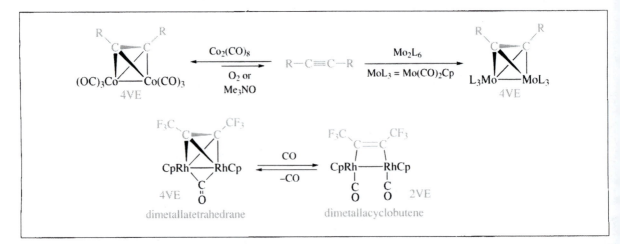

Figure 6.34 Binuclear alkyne coordination

If simple mononuclear alkyne coordination does not complete an 18VE count for the metal, then the possibility arises for further donation of the 2VE housed in the orthogonal π-bond (Figure 6.33). If the metal centre possesses empty orbitals of appropriate π-symmetry, then the alkyne effectively serves as a π-donor through this interaction. In valence bond terms (Figure 6.35) we may depict this in two ways: either as a π-donor or alternatively as a bis(alkylidene). Both depictions have merit and the latter is supported by the observation that the adoption of 4VE donation is accompanied by a decrease in metal–carbon bond lengths and an increase in the C–C bond length. Furthermore, 4VE alkyne coordination also results in a shift to low field of the alkyne ^{13}C NMR resonance, towards the region associated with metal–carbon multiple bonding. In general, 4VE alkyne coordination is more prevalent for metals of the earlier transition triads with <6 d-electrons, in particular d^4 octahedral complexes (i.e. t_{2g}^4) are very common. From Figure 6.36 it becomes apparent that in some instances it is not clear just how many electrons are provided by an individual alkyne, in which case a more detailed molecular orbital treatment is required.

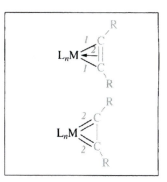

Figure 6.35 4VE alkyne coordination

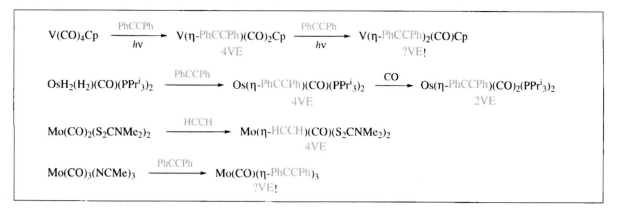

Figure 6.36 2VE vs. 4VE alkyne coordination

As with alkenes, coordination of an alkyne may activate it towards electrophilic attack, nucleophilic attack, insertion or ligand coupling reactions. Alkynes may also be metathesized by alkylidyne complexes (Chapter 5), and metallacyclobutadienes may occasionally be encountered in such processes. The arguments for choosing a particular metal centre to favour either nucleophilic attack or electrophilic attack for alkenes, alkylidenes and CO apply equally to alkyne complexes. A generalized scheme is depicted in Figure 4.12, whilst Figure 6.37 illustrates the specific case of nucleophilic attack at alkynes coordinated to the electrophilic 'Fe(CO)(PR$_3$)Cp' (R = Ph, OPh) cationic complex, for comparison with Figure 6.10. The products are generally σ-alkenyl complexes; however, some distinctions do arise. In contrast to alkene-derived alkyls, alkenyls which are obtained by nucleophilic attack at alkynes offer the possibility of interaction of the C=C double bond with the metal to provide σ,π-alkenyl ligands (3VE); this possibility may arise when 4VE alkyne coordination is involved, or for 2VE alkyne complexes in which one co-ligand is potentially labile.

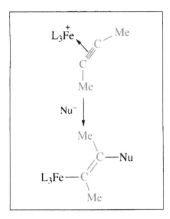

Figure 6.37 Nucleophilic attack at coordinated alkynes; FeL$_3$ = Fe(CO)(PR$_3$)Cp; R = Ph, OPh; Nu = Me, Ph, CN, CH(CO$_2$R)$_2$, OR, SR

Figure 6.38 Metal-mediated oligomerization of alkynes

By far the most overwhelming class of reaction for alkynes with transition metals is their ligand coupling reactions, including self-coupling processes (dimerization, oligomerization). This field requires a tome in its own right and so only some illustrative examples will be provided. Simple oxidative dimerization of *terminal* alkynes to provide symmetrical 1,3-diynes has long been routinely achieved using copper salts, and more recently by a variety of transition metal complexes. For internal alkynes, however, which lack mobile protons, other structures may be obtained. The conceptually simplest outcome would be cyclo-butadiene formation; however, since these molecules are anti-aromatic (4π), they are not generally liberated from the metal centre. Instead, stable complexes often arise (Figure 6.38). One generally accepted mechanism (*cf.* Figure 6.22) involves the intermediacy of metallacy-clopentadienes (which may sometimes be isolated). Reductive elimina-tion of the two M–C bonds generates the cyclobutadiene ligand, which remains coordinated. However, this intermediate may also react with a further equivalent of alkyne (*via* one of two possible mechanisms; inser-

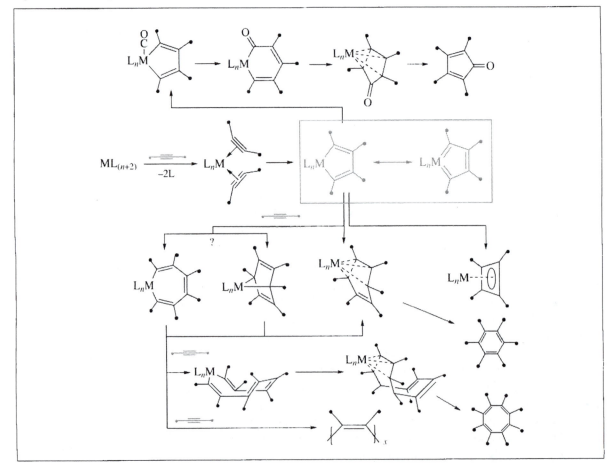

tion or [2 + 4] *pseudo*-Diels–Alder) to provide an arene. If this is eliminated from the metal centre (as is usually the case), the unsaturated complex may react with further alkyne in a catalytic synthesis of arenes. This principle has been extended to the cross coupling of alkynes and nitriles, leading to pyridine formation. If the metal centre involved also has carbonyl co-ligands, then the reaction may be diverted towards the formation of cyclopentadienones, tropones or quinones, and η^4, η^6 and η^2 complexes of these ligands have all been isolated.

A number of complexes react with internal alkynes to provide stable metallacyclopentadienes. These occasionally appear to have a 16VE metal centre; however, it should be noted that an alternative valence bond representation involving metal–carbon multiple bonding may be envisaged, in which case an analogy with pyrroles and thiophenes emerges (see Figure 7.19). These metallacycles, in particular those based on Group 4 metallocenes, can serve as precursors for a wide range of five-membered unsaturated heterocycles by treatment with suitable main-group element dihalides *via* electrophilic cleavage, whilst hydrolysis provides the corresponding butadiene (Figure 6.39).

The most effective and widely used reagents for alkyne trimerization are sources of the 14VE 'CoCp' fragment, *e.g.* CoL_2Cp (L = CO, PPh_3, C_2H_4).

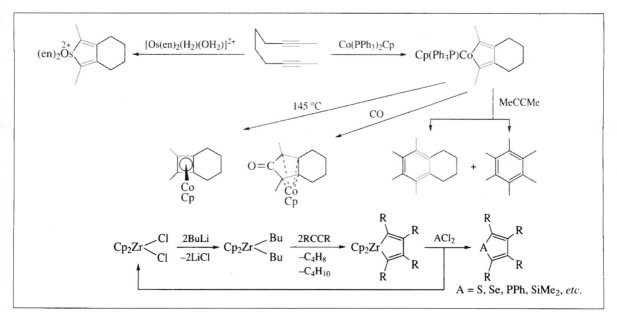

We have already mentioned cyclooctatetraene and will meet more examples of its complexes. This aspect of organometallic chemistry might never have evolved to its present state if chemists needed to rely on the laborious multi-step synthesis. Reppe (1948), however, found that a variety of nickel complexes [*e.g.* $Ni(CN)_2$, $Ni(acac)_2$, $Ni(H_2C=CHCN)_2$] could cyclotetramerize ethyne (acetylene) to provide this hydrocarbon in good yield (70%, 80–120 °C, 10–25 bar). Depending on the conditions,

Figure 6.39
Metallacyclopentadienes; en = 1,2-diaminoethane (ethylene-1,2-diamine)

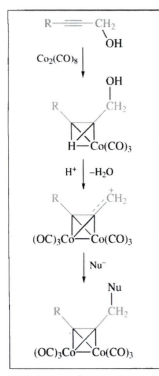

Figure 6.40 Propargylic substitution

ligands and choice of solvent, the reaction may be diverted to provide either simple dimerization (butenyne), trimerization (benzene) or polymerization (polyacetylene). Small amounts of styrene, phenylbutadiene and vinylcyclooctatetraene and traces of azulene and naphthalene (*via* cyclodecapentaene) are also observed. Numerous other metal catalysts have since been shown to access this type of chemistry.

The coupling of alkynes, alkenes and CO to provide cyclopentenones *via* the intermediacy of dicobaltatetrahedranes (Figure 6.34) is known as the Pauson–Khand reaction. The reaction of cobalt carbonyl with alkynes to provide these intermediates is virtually quantitative and, once formed, the initial alkyne may be substituted by more electronegative alkynes. The alkynes may themselves be modified whilst coordinated, in which case the dicobalt moiety might be viewed as a protecting group for the alkyne. Some of the most useful protocols include the convenient modification of coordinated propargylic alcohols (Figure 6.40). Abstraction of the hydroxide (Brønsted or Lewis acids) adjacent to the alkyne leads to a carbonium ion (comparable stability to Ph_3C^+), which may then be functionalized with a variety of nucleophiles. The reactions of these tetrahedranes with alkenes and CO provides cyclopentadienones, with good regioselectivity (CO adjacent to the bulkier alkyne substituent) in the C–C bond-forming stages. Chiral centres are generated if the alkene is prochiral, and the introduction of chiral co-ligands may lead to enantioselectivity. In some cases the reaction may be carried out using catalytic amounts of $Co_2(CO)_8$; however, it is more usual to isolate the alkyne complex intermediate (Figure 6.41).

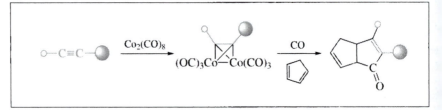

Figure 6.41 Pauson–Khand synthesis

7

η^n-C_nR_n Carbocyclic Polyene Ligands (n = 3–8)

Aims

By the end of this chapter you should have an appreciation of the organometallic chemistry of carbocyclic polyene ligands in both spectator and participatory roles. This will include:

- The bonding of representative examples and the implications for structures adopted
- Synthetic routes for introducing these ligands
- Typical reactivitiy profiles for such ligands
- Some illustrative applications of complexes of these ligands

7.1 Introduction

This chapter is concerned with planar carbocyclic polyene and polyenyl ligands of the form cyclo-C_nR_n (n = 3–8) in addition to the related odd-electron acyclic ligands η^n-C_nR_{n+2} (n = 3, 5). The material is arranged according to hapticity. However, before embarking on a systematic march from η^3 to η^8, centre stage must be given to ferrocene, Fe(η-$C_5H_5)_2$, arguably the most important synthetic compound of the 20th century. This remarkable organometallic compound was discovered at a time when the field of simple transition metal–carbon single bonds appeared to be a non-starter. At this time, new spectroscopic techniques that we now consider routine were slowly becoming available, and could be used in combination with (at the time painfully arduous) X-ray crystallography to test the emerging theories on metal–ligand bonding. For such a completely novel class of molecule, and an archetype of such elegant simplicity, to appear at such a scientifically voracious time explains why it captured the imagination of such a diverse group of chemists. In the intervening 50 years, only buckminsterfullerene (C_{60}) has

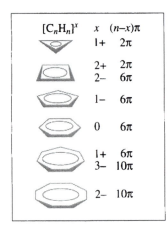

Figure 7.1 Aromatic carbocycles

When the remaining co-ligands differ markedly in their *trans* influences, *e.g.* oxo and imido co-ligands, the ring may tilt on the metal–centroid axis.

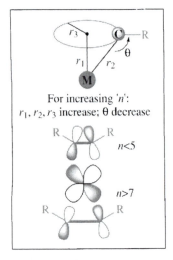

Figure 7.2 Geometrical implications for η^n-C_nR_n coordination

been so eagerly and widely embraced (including organometallic chemistry), for many of the same reasons.

For electron counting purposes and in deference to the principle of electroneutrality, we consider planar η^n-C_nR_n to serve as neutral ligands (Table 1.1). It is instructive, however, to consider the charges that the free molecules would require were they to satisfy the Huckel '$(4n + 2)\pi$' concept of aromaticity (Figure 7.1). Amongst these, only arenes ($n = 6$) would be expected to be equilateral, planar, neutral aromatic molecules. This perspective serves two purposes. Firstly, if (artificial) oxidation states need to be assigned, this *may* assist. Secondly, the relationship of charged species to related neutral molecules may provide retrosynthetic insight, *e.g.* the 6π cyclopentadienyl anion ($[C_5H_5]^-$) is easily generated by deprotonation ($-H^+$) of cyclopentadiene (C_5H_6), whilst hydride abstraction ($-H^-$) from cycloheptatriene (C_7H_8) provides the 6π tropylium cation ($[C_7H_7]^+$).

In addition to the carbocyclic ligands, we will also consider here the chemistry of allyl (C_3R_5) and pentadienyl (C_5R_7) ligands. These acyclic ligands might be considered as relatives of η^3-C_3R_3 and η^5-C_5R_5 ligands, although there is no circumannular delocalization, and the reactivity tends to be associated with the peripheral methylenic carbon atoms.

Before we consider each of these ligands in turn, it is useful to identify parallels, which run through the chapter and characterize all the η^n-C_nR_n ligands. These may be generalized as in Box 7.1.

Box 7.1 η^n-C_nR_n Ligands

- Coordination of the C_nR_n ring moderates the reactivity relative to the free ligand. The C_nR_n ligand in turn moderates the reactivity of the metal centre.
- For a symmetrical set of remaining co-ligands, the metal–centroid vector is usually normal to the C_n plane such that all n M–C bonds are of comparable length. The electronic delocalization throughout the ring is reflected in all C–C bonds being of comparable length (0.140–0.146 nm).
- The metal–centroid (r_1) and metal–carbon lengths (r_2) and ring radius (r_3) increase with the value of n as a simple requirement of geometry (Figure 7.2).
- For η^n-C_nR_n ligands the M–C–R angle increases from $n = 3$ to $n = 8$ such that the substituents are essentially coplanar with the C_n ring for $n = 5$ or 6. This pyramidalization of the ring carbons for $n = 3, 4, 7$ or 8 follows from the need to maximize the overlap between the metal valence orbitals and the orbitals

of the carbocyclic π-system. For small rings, bending the substituents away from the metal centre 'stretches' the orbital torroid on the metal-bound face. For large rings, bending of the substituents towards the metal contracts the orbital torroids (Figure 7.2).

- In solution the C_nR_n ring usually rotates very rapidly on the NMR and chemical timescales. Introduction of excess steric bulk into the ring substituents and/or low temperatures may slow this process.

- Coordination modes of reduced hapticity (η^{n-2x}, x = 1, 2, 3) may be observed, especially when the maximum hapticity would lead to >18VE. In such cases the metal may migrate between possible donor sites ('ring-whizzing') in solution (NMR evidence), although this is 'frozen out' in the solid state (IR, crystallography).

- The chemical shifts of single ^1H and ^{13}C NMR resonances for η^n-C_nH_n ligands depend on the nature of the metal centre; however, in general, the resonances move to lower field (i) as n increases and (ii) on descending a triad.

- Coordination of a metal centre to one or other face of the ring results in a loss of (planar) symmetry. If the C_nR_n (σ_h) plane is the *only* symmetry element present in the free ligand (*e.g.* in 1,2- or 1,3-XYC_6H_4), then the two faces will be diastereotopic. Accordingly, coordination of a metal to one or other of the pro-chiral faces will lead to enantiomers. If the metal centre itself is chiral, then diastereomers (different energy) arise upon coordination, perhaps selectively.

We will not consider here specific MO schemes for the various carbocyclic ligands since these are available in advanced texts. It is sufficient here to appreciate some general features of the frontier orbitals of η^n-C_nR_n ligands. An η^n-C_nR_n ligand provides nVE to the metal centre and presents n molecular orbitals (combinations of n p_z orbitals) for mixing with the nine valence orbitals of the metal centre. These may be of σ, π or (for n > 3) δ symmetry with respect to the vector joining the metal and ring centroid, allowing overlap with metal orbitals of σ (s, p_z, d_{z^2}), π (d_{xz}, d_{yz}, p_x, p_z) or δ (d_{xy}, $d_{x^2-y^2}$) symmetry. The actual occupancy of each of the resulting molecular orbitals (and their impact on reactivity/ stability) will depend on the number of valence electrons. This symmetry matching is illustrated for cyclobutadiene (Figure 7.3; *cf.* butadiene, Figure 6.26; see also Figure 7.23). The actual energetic ordering of the molecular orbitals which results from these combinations will depend on (i) the value of n, (ii) the nature of the metal centre, and (iii) the nature of the co-ligands which influence the relative energies of the metal valence orbitals.

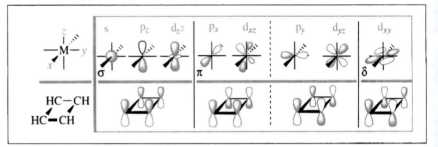

Figure 7.3 Symmetry considerations for metal valence orbitals and a representative η^n-C_nR_n carbocyclic ligand, C_4H_4

7.2   Allyls, η^3-C_3R_5

Allyl ligands (Figure 4.1) have already been encountered in one of their possible coordination modes: simple η^1 (1VE) coordination may be viewed as a sub-class of σ-alkyl ligands. The divergence arises, however, when the metal centre is able to accommodate a further pair of electrons (*i.e.* <18VE) from the double bond, thereby adopting a *trihapto* coordination. In valence bond terms (Figure 1.6), a contribution to stability can be envisaged from three canonical forms (both C–C bonds of similar length). In molecular orbital terms (Figure 7.4) the hydrocarbyl presents orbitals of both σ and π symmetry. The importance of the central molecular orbital (localized on the terminal carbon atoms) is manifest in these carbons having shorter metal–carbon bond lengths such that the C_3 plane is tilted, *e.g.* by 20° in the homoleptic complex $Ni(\eta$-$C_3H_5)_2$, accompanied by displacement of the allyl substituents from the C_3 plane. Cyclopropenyl ligands present a special case (Section 7.4).

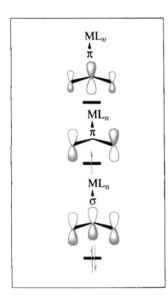

Figure 7.4 Allyl (C_3H_5) π molecular orbitals

The η^1 and η^3 coordination modes are often interconvertible *via* the addition or dissociation of co-ligands, which modifies the EAN requirements of the metal. Many allyl compounds have complex NMR spectra due to the various fluxional processes that may operate. For η^1-allyl ligands the possibility of transfer of the metal between the ends of the allyl group may occur, chemically equilibrating the two ends of the hydrocarbyl on the NMR timescale. For *trihapto* allyl coordination, rotation about the axis perpendicular to the C_3 plane may occur, in addition to reversible η^1–η^3 tautomerism. The intermediacy of η^1 coordination allows the possibility of rotation about the $C(sp^3)$–$C(sp^2)$ bond such that the *syn* and *anti* substituents may exchange sites (Figure 7.5). So long as these possibilities are appreciated, the chemical shifts for protons of η^3-allyl ligands fall into reasonably distinct ranges ($\delta_{meso} > \delta_{anti} > \delta_{syn}$), whilst the relative magnitudes of $^2J(H_{syn}$–$H_{anti}) < {}^3J(H_{syn}$–$H_{meso}) < {}^3J(H_{anti}$–$H_{meso})$ assist in making assignments. In the case of η^1-allyl

Since η^1–η^3 tautomerism involves an intermediate of reduced coordination number, the process may be dependent on the nucleophilicity of the solvent which may coordinate transiently.

ligands, the resonances fall into distinct aliphatic (M–CH$_2$) and vinylic (CH=CH$_2$) regions.

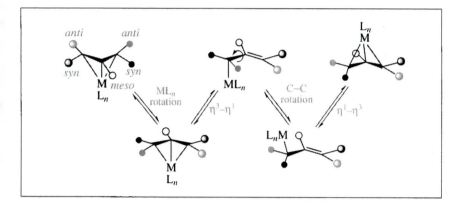

Figure 7.5 Fluxional processes for η^3-allyl ligands

Allyl complexes may be accessed *via* comparatively general synthetic routes typical of simple σ-alkyl ligands, *i.e.* transmetallation, oxidative addition of allyl halides or nucleophilic substitution of allyl halides by carbonyl metallates (Figure 4.6) and the hydrometallation of allenes or 1,3-butadienes (Figure 6.32). Furthermore, the rearrangement of alkenes bearing allylic hydrogen substituents has been noted (Figure 6.7). Allyl Grignard reagents (commercially available) are the most widely employed nucleophilic transfer reagents, especially for the preparation of homoleptic complexes (Figure 7.6).

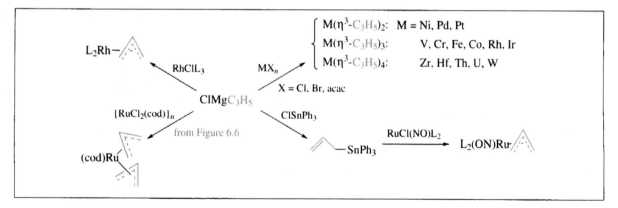

M(η^3-C$_3$H$_5$)$_2$: M = Ni, Pd, Pt
M(η^3-C$_3$H$_5$)$_3$: V, Cr, Fe, Co, Rh, Ir
M(η^3-C$_3$H$_5$)$_4$: Zr, Hf, Th, U, W

Figure 7.6 Synthesis of allyl complexes *via* transmetallation (nucleophilic organyl delivery, L = PPh$_3$)

Allyl halides (electrophilic allyl delivery) are effective both for oxidative addition and for reactions with nucleophilic carbonyl metallates (Figures 7.7, 7.8). In some cases, *e.g.* with Na[M(CO)$_5$] (M = Mn, Re), *monohapto* allyl intermediates may be isolated and then photochemically or thermally decarbonylated.

Whilst homoleptic alkyls are generally highly reactive and kinetically unstable, the possibility of η^3-coordination has made the isolation of

The benzyl ligand CH$_2$Ph may also be considered as a special case of allyl coordination wherein one double bond of the arene becomes associated with the metal centre (Figure 4.1).

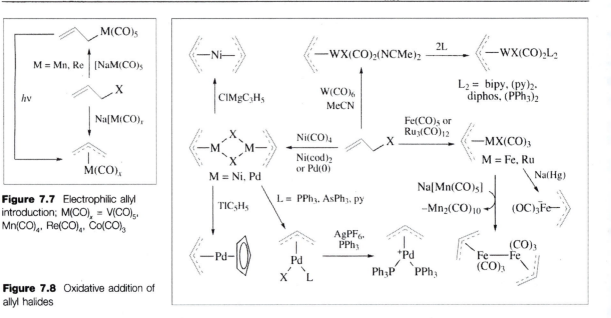

Figure 7.7 Electrophilic allyl introduction; M(CO)$_x$ = V(CO)$_5$, Mn(CO)$_4$, Re(CO)$_4$, Co(CO)$_3$

Figure 7.8 Oxidative addition of allyl halides

many homoleptic allyl complexes comparatively routine (Figure 7.6). Poly(allyl) complexes are useful starting materials owing to their solubility in many solvents, comparative stability and high reactivity, *e.g.* towards hydrogenolysis, protonolysis or reductive elimination, providing a route to unsaturated metal species.

The reactivity of allyl ligands recalls aspects of both σ-organyl (Chapter 4) and also π-alkene (Chapter 6) chemistry. In a σ-organyl context, the case of electrophilic cleavage deserves special comment. In addition to the metal centre and the M–C bond, the pendant C=C double bond may also be attacked by electrophiles. However, the initial site of attack may not always be unequivocally evident from the nature of the alkene product (Figure 7.9a). The chemistry of η^3-allyl complexes is strongly dependent on the nature of the metal centre. Thus the reactions of nickel(II) complexes are dominated by the organyl ligand displaying nucleophilic character (Ni PE = 1.9), which may be exploited in coupling reactions with alkyl, aryl, vinyl or allyl halides (many of which fail in the case of allyl Grignard reagents) (Figure 7.9b). Intramolecular variants may be devised if both the allylic halide and the subsequent electrophile are contained within the same substrate (Figure 7.9c).

Whilst electrophilic attack is commonly observed for η^1-allyl ligands, the reactivity of η^3-allyl ligands towards nucleophiles has been intensely studied because the enormous synthetic utility this offers for allylic functionalization (allylic alkylation). The metal of choice for such applications is generally palladium (PE = 2.2), which may be employed in numerous catalytic processes. The versatility of palladium arises from the general tolerance of late transition metal centres to polar functional

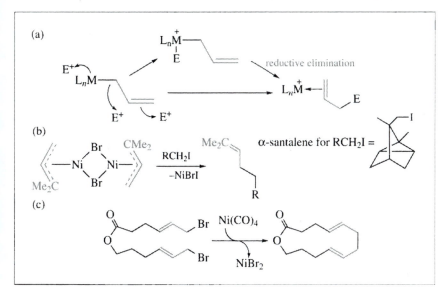

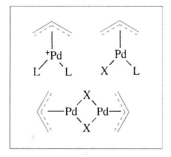

Figure 7.9 Electrophilic attack at σ^1- and η^3-allyl ligands

Figure 7.10 Allyl palladium complexes

groups, minimizing the requirement for protection/deprotection proto-cols. The key step is the facile oxidative addition of the allyl–X bond (X = halide, OAc, OPh, OH, NR_3^+, SO_2Ph) to a palladium(0) precursor to provide (depending on the ligands present and choice of solvent) η^3-allyl complexes of the forms shown in Figure 7.10.

It is not always clear which one (or more) of these forms operates, although the cationic species might appear to be most prone to nucle-ophilic attack. We have already seen that coordination of alkenes to diva-lent palladium activates them towards nucleophilic attack. In a similar manner, a wide range of nucleophiles may be introduced in a step which, in generating a new alkene, also reduces the palladium (II → 0) for intro-duction into the next cycle (Figure 7.11).

A wide range of nucleophilic functional groups may be introduced; however, of particular use are C–C bond-forming reactions employing stabilized carbanions (malonates, dithianyls, *etc.*). Intramolecular appli-cations of this reaction are particularly efficient, allowing the construc-tion of 3- to 11-membered rings. For most nucleophiles (including stabilized carbanions), direct attack occurs at the allyl ligand on the face opposite to the metal. Since both the oxidative addition and the subse-quent *exo* nucleophilic attack involve inversion of configuration at car-bon, the net result (double inversion) is overall retention of chirality at the carbon. Alternatively, the nucleophile may attack the metal centre directly, followed by reductive elimination (proceeds with retention), such that the overall process results in stereochemical inversion at carbon. In the case of carbon nucleophiles, this is favoured by the use of organometallics of the main group elements (transmetallation), *e.g.* stannanes or (less commonly) Grignard reagents. This process finds

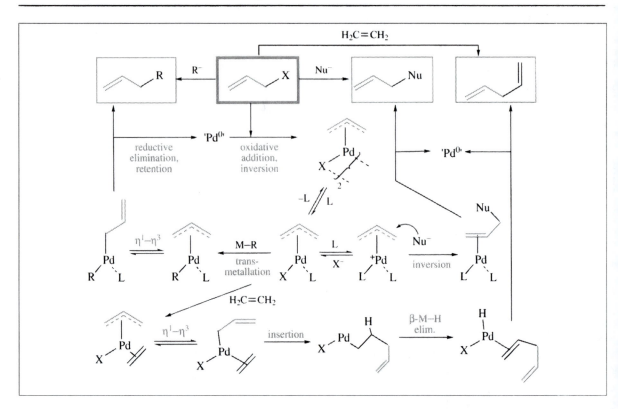

Figure 7.11 Generalized
scheme for palladium-catalysed
allylic functionalization

Whilst the catalysis begins with
Pd⁰, the combination of Pdᴵᴵ with
an alkene and nucleophile results
in reduction to Pd⁰. Accordingly,
the catalyst may also be
introduced in the form of a
divalent palladium complex
(typically and conveniently more
air stable).

parallels with the palladium-catalysed nucleophilic substitution of aryl halides (Figure 4.10). If an alkene is introduced in place of the nucleophile, then coordination to the metal centre may be followed by insertion of the alkene into the metal–allyl bond (possibly *via* an η^1-tautomer). The resulting alkyl may undergo β-M–H elimination to provide a 1,4-diene. The amazing utility of this technology is beyond this discussion; however, an illustrative selection of organic syntheses employing this technology is presented in Figure 7.12.

7.3   Pentadienyls, η^5-C_5R_7

Pentadienyls may be considered as vinylogues of allyl complexes and accordingly parallels exist, both in their synthesis and reactivity. These are most widely prepared by transmetallation from lithium or Grignard reagents, nucleophilic attack at pentadienyl halides or *via* hydride abstraction from η^4-pentadiene complexes (Figure 7.13).

A special case involves complexes of cyclohexadienyl ligands, which may result from the addition of nucleophiles to η^6-arene complexes (Section 7.7) or hydride abstraction from complexes of readily available cyclohexadienes (Birch reduction of arenes) (Figure 7.14). In the latter case, it is within the chemistry of iron that such complexes find the widest

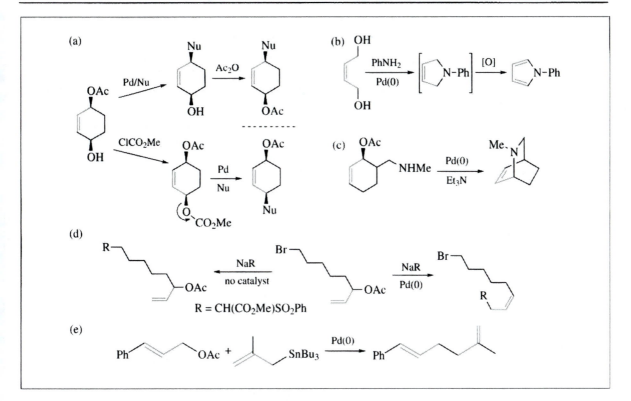

Figure 7.12 (Above) Illustrative examples of palladium-catalysed allylic functionalization

Figure 7.13 Synthesis of pentadienyl complexes

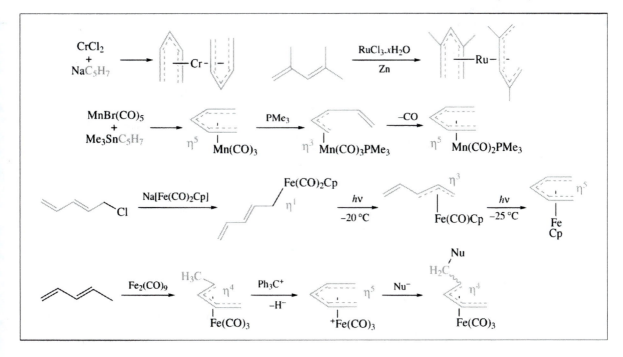

application in organic synthesis. By microreversibility, the abstraction of hydride from a pentadiene ligand suggests that the point of abstraction (the pentadienyl terminus carbon) will be prone to nucleophilic attack (Figure 7.14). Bis(pentadienyl) complexes have been prepared (typically by transmetallation) and their reactivity compared to those of metallocenes (Section 7.6). From this it may be concluded that the pentadienyl ligand is typically more likely to take part in ligand-based reactions than are cyclopentadienyl ligands, and also that it is more prone to reduction in hapticity (Figure 7.14).

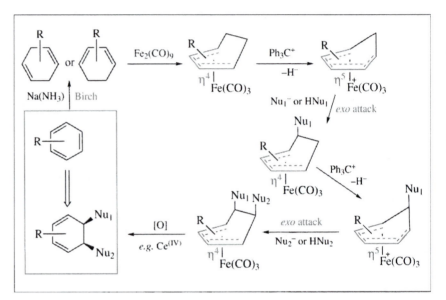

Figure 7.14 Synthetic applications of cyclohexadienyl iron complexes; HNu_x, Nu_x^- = H^-, OH^-, OR^-, NH_3, NH_2R, NHR_2, R^- (R–M or stabilized carbanion)

7.4   Cyclopropenyls, η^3-C$_3$R$_3$ (Metallatetrahedranes)

The cyclopropenium cation $[C_3H_3]^+$ is the simplest aromatic (2π) carbocyclic system. However, having only three p-orbitals with which to construct (π) frontier orbitals (Figure 7.15), there is no orbital of δ symmetry. The chemistry of metal cyclopropenyls is far less studied than that of the acyclic allyl ligands discussed previously. Cyclopropenyl halides readily ionize to provide aromatic (2π) cyclopropenium salts (isolated with non-coordinating anions). Alternatively, a hydride may be abstracted from the sp^3 carbon of cyclopropenes (e.g. with $[Ph_3C]BF_4$) to provide isolable cyclopropenium salts, amongst which the triphenyl derivative has been the most widely studied in an organometallic context. These and the halides serve as the most common means for introducing the cyclopropenyl ligand, via oxidative addition or nucleophilic attack by carbonyl metallates (Figure 7.16).

In some cases, η^1-cyclopropenyl complexes may be isolated as intermediates, whilst in others, migration to a carbonyl ligand may be

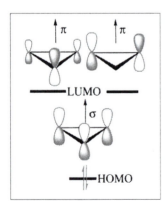

Figure 7.15 Frontier molecular orbitals for $[C_3H_3]^+$

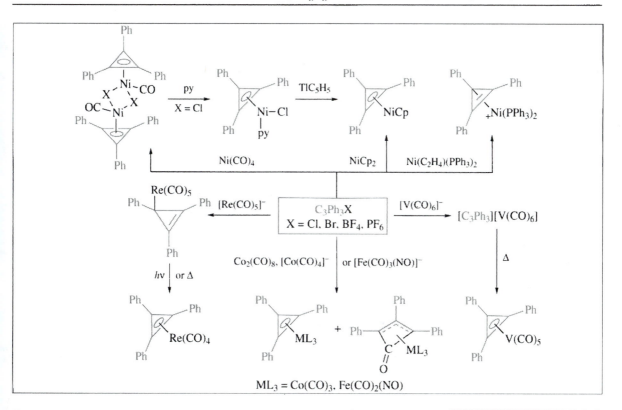

Figure 7.16 Synthesis of cyclopropenium complexes

followed by insertion of the CO group into the (strained) carbocycle, resulting in the formation of η^3-cyclobutenonyl ligands. Two special cases in which cyclopropenes do not lead to cyclopropyl complexes are worthy of note: 3,3-dihalocyclopropenes have been used for the synthesis of cyclopropenylidenes, whilst some cyclopropenes may be ring opened to provide alkenyl carbenes (Figure 7.17).

Although many conventional and symmetric η^3-cyclopropenyl complexes are known, in many cases quite dramatic distortion from ideal local C_{3v} MC$_3$ geometry may occur. The reactions of [C$_3$Ph$_3$]BF$_4$ with Pt(η-C$_2$H$_4$)(PPh$_3$)$_2$ and IrCl(CO)(PMe$_3$)$_2$ are of interest in that the products do not contain symmetrical η^3-C$_3$Ph$_3$ ligands. Instead, a distorted tetrahedrane arises in the former and a metallacycle (iridacyclobutadiene) in the latter, wherein the iridium is inserted into one C–C bond (Figure 7.18). This product recalls the formation of such metallacycles *via* the cycloaddition of alkynes to high oxidation state Group 6 alkylidynes (Figure 5.44), and even these metallacycles are occasionally found to be non-planar.

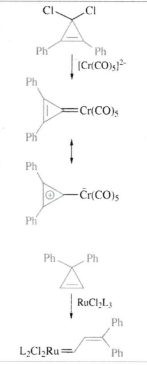

Figure 7.17 Cyclopropene-derived carbenes; L = PPh$_3$

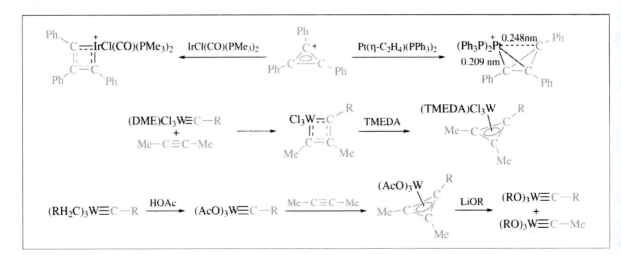

Figure 7.18 Cyclopropenyl vs. metallacyclobutadiene coordination; R = Bu^t; Ac = COMe

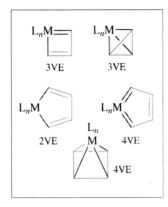

Figure 7.19 Metallacyclic vs. polyhapto coordination

Ni(η-C$_4$Ph$_4$)$_2$ is notable as the first metal sandwich based on two C$_4$ rings [completing the series Cr(η-C$_6$R$_6$)$_2$, Fe(η-C$_5$R$_5$)$_2$ and Ni(η-C$_4$R$_4$)$_2$] and also because its isolation (m.p. 404 °C!) defied theoretical calculations suggesting that d^{10} complexes of cyclobutadienes would not be

It is perhaps premature to attempt to delineate the factors that determine whether symmetrical or distorted metallacyclobutadiene or η^3-cycopropenyl coordination is observed, given the comparitive sparsity of directly comparable examples, and the observation that tautomerism appears to operate between coordination modes in some cases. A similar situation arises for cyclobutadiene coordination vs. metallacyclopentadiene formation (Figures 6.38, 6.39, 7.19). In both cases the metallacycles are important intermediates in catalytic manifolds (alkyne metathesis and oligomerization, respectively) and in both cases the *polyhapto* variant represents a tangent to the productive catalytic cycle, formation of which may be reversible or in some cases may lead to termination.

7.5   Cyclobutadienes, η^4-C$_4$R$_4$

As early as 1956 it was suggested that the anti-aromatic ($4\pi \neq 4n + 2$) molecule cyclobutadiene (Figure 7.3) might be stabilized by coordination to a transition metal. This, coupled with the known ability of the Fe(CO)$_3$ unit to stabilize butadiene coordination, and the previous isolation of complexed substituted cyclobutadienes (Figure 7.20), led Pettit to design and isolate Fe(CO)$_3$(η-C$_4$H$_4$) from the reaction of dichlorocyclobutene with Fe$_2$(CO)$_9$. In this reaction the iron carbonyl acts as both a reductant and a trap. More generally, the majority of cyclobutadiene complexes arise from alkyne coupling, prior to or as a result of coordination (Figure 7.20). Many examples of both metallacyclopentadienes and cyclobutadiene complexes (Figure 7.19) are known. In some cases the former lies synthetically *en route* to the latter *via* reductive elimination. In other cases this final step is not easily induced. One case where ring closure (reductive elimination) does occur is in the formation of Ni(η-C$_4$Ph$_4$)$_2$ (Figure 7.20).

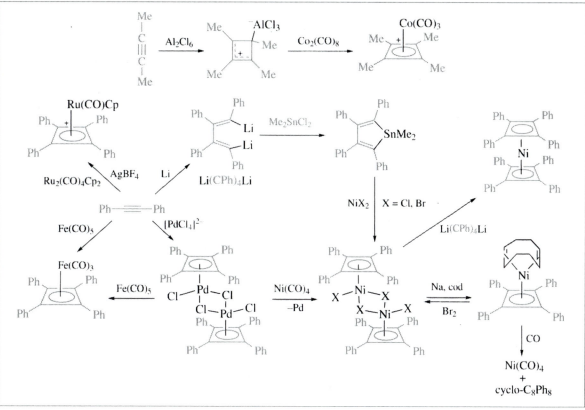

Figure 7.20 Synthesis of cyclobutadiene complexes

On reaching η^4 in our march, aromatic character begins to become manifest in the reactivity of the coordinated carbocycle. Figure 7.21 illustrates a range of reactions of Pettit's complex, including those involving functionalization of the coordinated ring (*e.g.* Friedel–Crafts acylation) and the reactions of free cyclobutadiene which may be liberated by oxidation (Ce^{IV}) in the presence of suitable trapping agents.

7.6   Cyclopentadienyls, η^5-C_5R_5

Complexes of the cyclopentadienyl ligand fall loosely into three classes: metallocenes, bent metallocenes and half-sandwich complexes (Figure 7.22). For metallocenes, interest focuses on the chemistry of the cyclopentadienyl rings, whilst for bent metallocenes and half-sandwich complexes the cyclopentadienyl ligands *usually* serve as innocent spectator ligands. A special case of bent metallocenes involves those wherein the two rings are joined by a bridge (*ansa*-metallocenes, metallocenophanes), which may impart steric (and resulting electronic) constraints upon the complex geometry.

A detailed discussion of the molecular orbital scheme for linear sand-

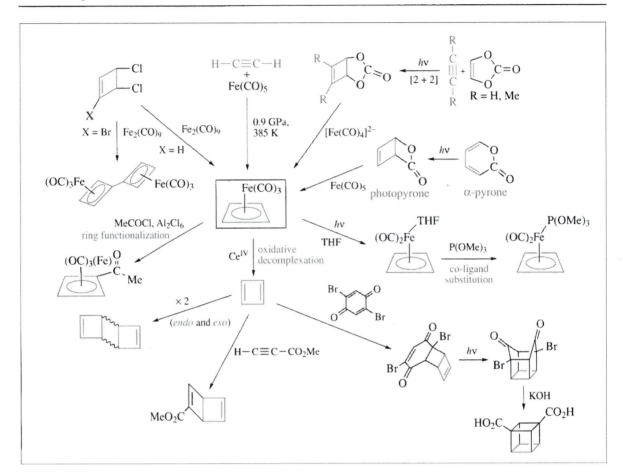

Figure 7.21 Syntheses and selected reactions of Pettit's complex, Fe(CO)$_3$(η-C$_4$H$_4$)

wich complexes will be deferred until we consider Cr(η-C$_6$H$_6$)$_2$. This is justified on three grounds. Firstly, the molecular orbitals of benzene (0–3 nodal planes bisecting a six-membered ring) are more familiar and easily visualized than those of the cyclopentadienyl anion (0–2 nodal planes superimposed on a five-membered ring). Secondly, the orbital symmetries and principles illustrated apply to both Fe(η-C$_5$H$_5$)$_2$ and Cr(η-C$_6$H$_6$)$_2$. Finally, so long as we appreciate the limitations of a crystal field approximation, i.e. neglect of covalency, the actual ordering of the resultant frontier orbitals is similar to that arising from a molecular orbital approach. This is not an acceptable approximation, but will suffice in the interim. The immediate conclusion from Figure 7.23 is that the frontier orbitals comprise a (closely spaced) set of a$_{1g}$ (d$_{z^2}$) and degenerate e$_{2g}$ (d$_{x^2-y^2}$, d$_{xy}$) orbitals which lie below two degenerate e$_{1g}$* (d$_{xz}$, d$_{yz}$) orbitals which are destabilized as a result of pointing towards the C$_5$H$_5$ orbital torroids. The most stable situation would therefore appear to be for an 18VE complex with vacant e$_{1g}$* orbitals. This is indeed the case, with the d^6 metallocenes MCp$_2$ (M = Fe, Ru, Os, Co$^+$; Cp = η-C$_5$H$_5$)

being the most stable and inert. However, the close spacing of the a_{1g} and e_{2g} orbitals means that 3d metallocenes which have these only partially filled are also comparatively stable (*e.g.* M = Cr, V, Mn), although, being electron deficient, they show a tendency to coordinate ligands [*e.g.* M(CO)Cp$_2$ (M = V, Cr)]. Similarly (paramagnetic) MCp$_2$ (M = Co, Ni) are also thermally stable despite population of the e_{1g}* orbitals.

For an 18VE metallocene the three e_{2g} and a_{1g}* orbitals are primarily metal based and weakly M–C antibonding. If we imagine bending the metallocene, the effect is to 'squeeze' these orbitals out between the rings, making them available for interaction with further ligands in the plane between the two Cp rings (Figure 7.22). The occupancy of these orbitals will depend on the overall valence electron count. Thus a d^0 bent metallocene is Lewis acidic; for TiCl$_2$Cp$_2$, two orbitals become involved in bonding to chloride ligands whilst the third acts as a Lewis acid. For a d^2 or d^4 bent metallocene, however, one of these orbitals will be occupied and may be attacked by electrophiles, *e.g.* WH$_2$Cp$_2$ (d^2) and ReHCp$_2$ (d^4) which are protonated to form [WH$_3$Cp$_2$]$^+$ and [ReH$_2$Cp$_2$]$^+$, whilst TaH$_3$Cp$_2$ (d^0) is not. The implications of this rehybridization are that with only three coplanar orbitals available, a maximum of three

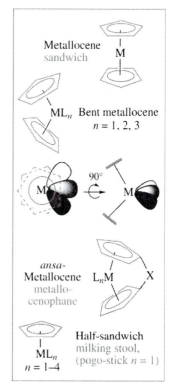

Figure 7.22 Cyclopentadienyl complexes

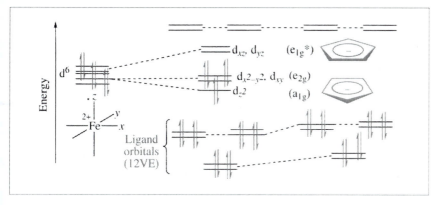

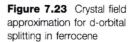

Figure 7.23 Crystal field approximation for d-orbital splitting in ferrocene

covalent coplanar metal–ligand bonds may be formed. Effective metal–ligand π-interactions are limited to these three orbitals, which lie in the plane between the two rings, *e.g.* the loss of this interaction contributes to orientation and the high rotation barrier for the carbene ligand in Ta(=CHPh)(CH$_2$Ph)Cp$_2$ (Table 1.3).

7.6.1 Ferrocene

The chemistry in this chapter originates from Pauson's discovery of bis(cyclopentadienyl)iron(II). Pauson had attempted the synthesis of the 10π aromatic hydrocarbon fulvalene *via* an oxidative coupling/dehydrogenation of the cyclopentadienyl anion (Figure 7.24). However, Pauson

Many years later it was found that Na[C₅H₅]DME could be oxidatively coupled to provide dihydrofulvalene, using iodine as the oxidant, and that this provided access to fulvalene complexes.

chose ferric [iron(III)] chloride (commonly employed for oxidative biaryl coupling) and obtained a highly stable, sublimable, diamagnetic orange compound of composition $C_{10}H_{10}Fe$ which he formulated as $Fe(\sigma\text{-}C_5H_5)_2$.

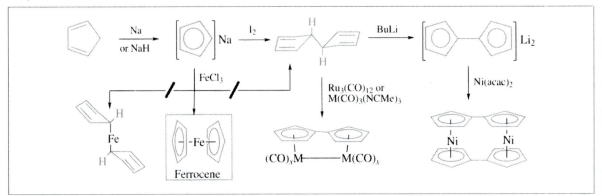

Figure 7.24 Synthesis of ferrocene and fulvalene complexes; $M(CO)_x$ = $Ru(CO)_2$, $Cr(CO)_3$, $Mo(CO)_3$, $W(CO)_3$

Fischer and Wilkinson were jointly awarded the Nobel Prize in 1973.

The apparent instability of metal–carbon single bonds, which was prematurely presumed at the time, led Wilkinson and Fischer to question this formulation. They proposed and confirmed the now-familiar sandwich structure wherein the metal resides between parallel planar C_5H_5 rings.

This revolutionary formulation was supported by both the physical properties (zero dipole moment, diamagnetism, IR and X-ray crystallography) and also the chemical reactivity. Amongst the first of these to be investigated were electrophilic aromatic substitution reactions (*e.g.* Friedel–Crafts acylation). Not only did ferrocene undergo electrophilic aromatic substitution, it did so with far greater activity than simple arenes (*ca.* 3×10^6 as fast as C_6H_6). This pseudo-aromatic behaviour led to the trivial name 'ferrocene' and later to the more general term 'metallocene' as it became clear that ferrocene was by no means unique in forming bis(cyclopentadienyl) complexes. Figure 7.25 presents a range of key ferrocenyl derivatives which arise from electrophilic attack at one of the C_5H_5 rings; however, it is an illustrative over-simplification. In many cases, attack at both rings may occur, requiring careful control of conditions. In general, introduction of an electron-withdrawing substituent will deactivate a cyclopentadienyl ring to further electrophilic attack, thereby directing attack to the second ring. This is illustrated by the acid-catalysed H/D exchange of variously substituted ferrocenes $Fe(\eta\text{-}C_5H_5)(\eta\text{-}C_5H_4R)$, the rate of which is found to decrease in the order $R = OMe > alkyl > H > Cl$, CO_2Me. Thus acetylation of acetylferrocene is approximately 10^4 times slower than for ferrocene, and produces a 10:1 mixture of the 1,1'- and 1,2-isomers (Figure 7.26). Conversely, Friedel–Crafts *alkylation* results in a more electron-rich alkyl-substituted

The deactivation by an acyl substituent is further increased since under Friedel–Crafts conditions the acyl coordinates to the Lewis acid (AlCl₃).

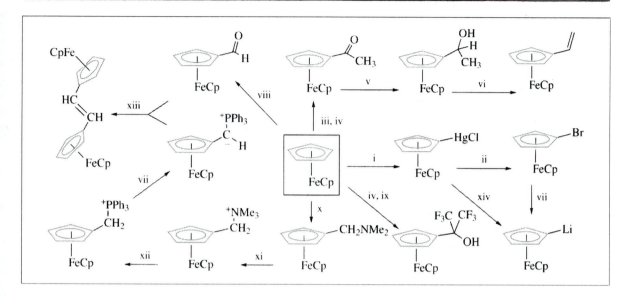

ring, which is therefore more prone to attack, and hence polyalkylation often occurs with little selectivity.

There is continuing debate about the role of the metal centre in the very facile electrophilic substitution reactions of ferrocene and related metallocenes. The Group 8 metallocenes are reversibly protonated at the metal by strong acids (Table 2.3), indicating that the metal centre has some nucleophilicity. The metallocene equivalent of the Wheland intermediate in conventional electrophilic substitution would be a cationic (16VE) cyclopentadiene complex that could in principle have the introduced electrophile *exo* or *endo* with respect to the iron. If this intermediate arose from initial attack at the metal centre followed by migration to the ring, the *endo* intermediate would result (although this could in principle also arise from direct *endo* ring attack). Figure 7.27 illustrates both mechanistic pathways in addition to results which point towards preferential direct *exo* attack. The two metallocene carboxylic acids shown were designed to only allow either *endo* or *exo* intramolecular attack (activated by *O*-trifluoroacylation). The metallocene capable of *exo* attack reacts faster than that for which only direct *endo* attack or attack at the metal centre is possible, *i.e.* denied access to the metal centre does *not* retard the rate. This argument assumes that electrophilic attack is the rate-limiting step and does not consider the possible participation of the metal centre in the subsequent deprotonation, *e.g. via* initial transfer to the metal centre.

Figure 7.27 is mechanistically most relevant to the functionalization of d^6 diamagnetic Group 8 metallocenes. A more general picture of the reactivity of metal-bound cyclopentadienyl ligands towards nucleophiles and radicals is presented in Figure 7.28, which takes into account

Figure 7.25 Aromatic character of ferrocene expressed in electrophilic ring functionalization. (i) Hg(O$_2$CMe)$_2$, LiCl; (ii) *N*-bromosuccinimide; (iii) ClC(=O)Me (Friedel–Crafts); (iv) Al$_2$Cl$_6$; (v) NaBH$_4$; (vi) H$^+$; (vii) BuLi; (viii) PhNMeCHO, POCl$_3$ (Vilsmeier); (ix) O=C(CF$_3$)$_2$; (x) H$_2$C(NMe$_2$)$_2$, MeCO$_2$H, H$_3$PO$_4$ (Mannich); (xi) MeI; (xii) PPh$_3$; (xiii) Wittig; (xiv) MeLi

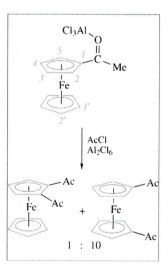

Figure 7.26 Selectivity in Friedel–Crafts acylation of ferrocene

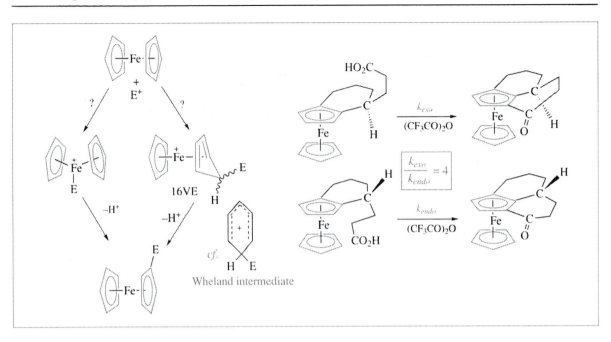

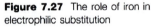

Figure 7.27 The role of iron in electrophilic substitution

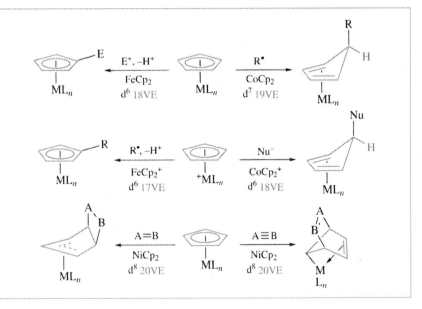

Figure 7.28 General reactivity of MCp$_2$ complexes

the effect of varying the nature (charge, nVE) of the metal centre, and the drive towards adherence to the EAN rule. In complexes where the dative capacity of the ligand exceeds the metal's requirements ($n > 18$), ring slippage may occur transiently, allowing the resulting pendant double bond to enter into cycloaddition chemistry, a characteristic reaction of 20VE nickelocene.

A complication in the reaction of ferrocene with electrophiles arises from the ease of one-electron oxidation of the metal centre by strong oxidants (Cl$_2$, NO$^+$, NO$_2^+$, *etc.*), to generate green/blue ferrocenium salts. The majority of reactions of ferrocene involve electrophilic attack, to which the d^5 ferrocenium cation is deactivated. This lack of reactivity, coupled with the solubility of ferrocene and ferrocenium salts, makes the [FeCp$_2$]$^{n+}$ (n = 0, 1) redox couple an ideal, chemically inert, internal reference for electrochemical experiments, and also as an electron carrier. This has been exploited in the use of the ferrocene/ferrocenium redox couple in redox 'biosensors' used for monitoring blood glucose levels in diabetics and in industrial fermentation monitoring. The metallocene acts as an electron carrier between the glucose oxidase enzyme (which specifically recognizes glucose) and a suitable electrode (Figure 7.29).

Given the remarkable stability of ferrocene, its chemistry might have been a *cul-de-sac* were it not for the high reactivity associated with the C$_5$H$_5$ rings, which allow very easy functionalization, using modifications of well-established protocols for simple aromatic compounds. Figure 7.25 showed reactions with electrophiles whilst Figure 7.30 illustrates a range of compounds which arise from σ-organometallic derivatives. These are far too numerous to cover here and for illustrative purposes a selection has been made which lead to ferrocenyl-based ligands. The key starting point in these syntheses involves ferrocene lithiation. Unfortunately, ferrocene reacts with BuLi to provide a mixture of mono-

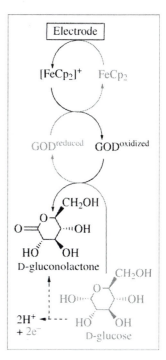

Figure 7.29 Ferrocene/glucose oxidase (GOD) based glucose sensor (current flow: red arrows)

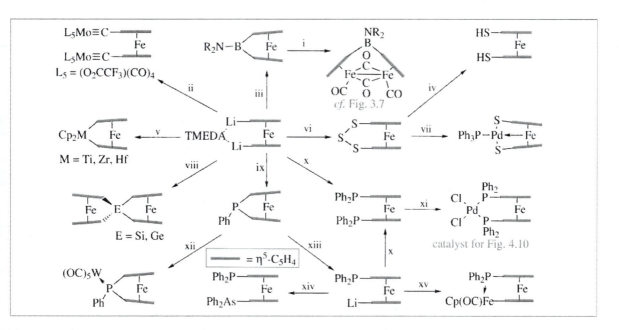

Figure 7.30 Ferrocenophanes and ferrocenyl ligands derived from dilithioferrocene. (i) Fe(CO)$_5$; (ii) Mo(CO)$_6$, (CF$_3$CO)$_2$O; (iii) Cl$_2$BNR$_2$; (iv) LiAlH$_4$; (v) MCl$_2$Cp$_2$; (vi) S$_8$; (vii) Pd(PPh$_3$)$_4$; (viii) ECl$_4$; (ix) Cl$_2$PPh; (x) ClPPh$_2$; (xi) PdCl$_2$(NCPh)$_2$; (xii) W(CO)$_5$(THF); (xiii) PhLi; (xiv) ClAsPh$_2$; (xv) FeCl(CO)$_2$Cp

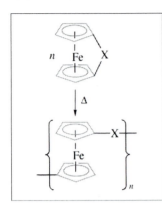

Figure 7.31 Ring-opening polymerization of metallocenophanes; X = PR, SiR$_2$, BR

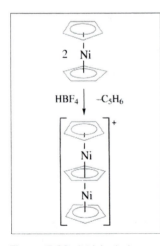

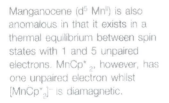

Figure 7.32 A triple decker

and 1,1'-dilithioferrocene, which is of limited use as a reagent (mixture). 1,1'-Dilithioferrocene, however, is formed in high yield from the combination of BuLi and TMEDA. Pure monolithioferrocene is obtained *via* the initial mercuration of ferrocene (Figure 7.25; a common reaction of arenes) by Hg(O$_2$CMe)$_2$/LiCl which, as in the case of acylation, provides a mixture of mono- and dimercurated compounds. These are air stable and easily separated. Amongst the derivatives shown in the figure are examples from a class of compounds known as metallocenophanes (*ansa*-metallocenes), *i.e.* compounds where the two rings are joined by a bridging unit (SiMe$_2$, PPh, BNMe$_2$). The rings in these molecules are not parallel, reflecting a degree of introduced strain which may be alleviated by thermolytic (or metal-catalysed) ring-opening polymerization to provide ferrocene-diyl polymers (Figure 7.31).

In addition to the many functionalized cyclopentadienyl ligands shown in Figures 7.23 and 7.30, a vast array of possible variants are accessible *via* modification of free cyclopentadienes. This is fortunate, in that ferrocene is unique in its versatility towards functionalization. The majority of other metallocenes, once constructed, are not as tolerant of such subsequent transformations.

7.6.2 Other Metallocenes

Ferrocene is somewhat special because of the position of iron(II) in the Periodic Table. Although the metals V, Cr, Mn, Fe (Ru, Os), Co and Ni all form thermally stable bis(cyclopentadienyl)s, it is only those of Group 8 which are diamagnetic. In solution, all of these others are air sensitive and mononuclear; however, in the solid state, manganocene adopts a chain structure with bridging C$_5$H$_5$ groups. Much of the chemistry of ferrocene is reflected in those of ruthenocene and osmocene, which, however, are less mature for reasons of cost and non-trivial or low-yielding synthetic routes. On moving away from iron, significant differences arise: to the right, the neutral metallocenes exceed 18VE and are easily oxidized, such that the cobaltocenium ion (18VE) is less reactive than cobaltocene (19VE). Whilst ferrocene is stable towards protonation, the protonation of nickelocene (20VE) results in loss of one C$_5$H$_6$ and formation of a triple-decker complex in which two nickels are sandwiched between three C$_5$H$_5$ rings (Figure 7.32). Although this was the first triple-decker complex, a number of others have now been prepared.

Although most of the 3d metals form stable metallocenes which contravene the 18-electron rule, titanium and many of the 4d and 5d metals do not tolerate this discomfort, but rather enter into reactions of the C$_5$H$_5$ rings which allow the metal to reach an 18VE count (Figure 7.33).

The chemistry of metallocenes may be divided into that associated

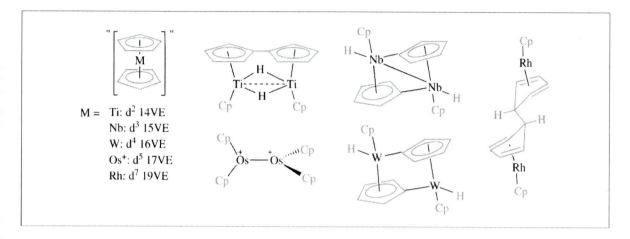

$$M = \quad Ti: d^2 \; 14VE$$
$$Nb: d^3 \; 15VE$$
$$W: d^4 \; 16VE$$
$$Os^+: d^5 \; 17VE$$
$$Rh: d^7 \; 19VE$$

with the hydrocarbon rings and that associated with the metal centre, wherein the C$_5$R$_5$ ring serves as a spectator ligand. Ring chemistry is best illustrated by the reactions of ferrocene itself (see above), and many of these extend to other metallocene or monocyclopentadienyl complexes so long as the co-ligands remain innocent. The basis for the immense number of such compounds is that when an η^n-C$_n$R$_n$ ligand coordinates to a metal centre, up to three mutually *cis* coordination sites are blocked, leaving the remaining sites for co-ligands which are mutually adjacent and well disposed for studying their interactions and modifications. One of the most important classes of such complexes is the cyclopentadienyl metal carbonyls (Figures 3.7–3.9, Table 3.4). Since the chemistry of these complexes focuses primarily on the other ligands, no more will be said here.

Figure 7.33 Elusive metallocenes

The trivial names **half-sandwich**, **piano-stool**, **milking stool** and **pogo-stick** are given to complexes which have only one cyclopentadienyl (or other η^n-C$_n$R$_n$) ligand.

7.7   Arenes, η^6-C$_6$R$_6$

The correct identification of the sandwich structure of ferrocene led Fischer to consider the possibility of arenes acting as *hexahapto* 6VE ligands. By simple arithmetic, a neutral bis(arene) sandwich complex of a zerovalent Group 6 element, *e.g.* chromium (Figure 6.71), was anticipated, a line of reasoning which led Fischer to develop the synthesis of dibenzenechromium. His approach resulted in the comparatively general 'Fischer–Hafner' synthesis (1955), which is applicable to many metals and arenes (devoid of Lewis-basic substituents, Figure 7.34).

The approach involves the reaction of a mid-valent metal halide with a mixture of arene, aluminium chloride (Lewis acidic halide abstractor) and aluminium powder (mild reductant). The charge on the resulting bis(arene) sandwich ($n+$) is generally dictated by the nature of the metal and EAN requirements, although once isolated, further reduction may be possible with stronger reductants. The approach can also be applied

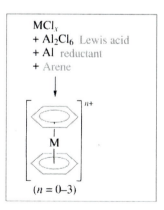

MCl$_x$
+ Al$_2$Cl$_6$ Lewis acid
+ Al reductant
+ Arene

(n = 0–3)

Figure 7.34 Fischer–Hafner synthesis; M = V, Cr, Mo, W, Tc, Re, Fe, Ru, Os, Co, Rh, Ir,

The use of a strong Lewis acid (Al_2Cl_6) precludes the use of arenes with nucleophilic functional groups (anilines, benzoate esters, etc.) and may also lead to the rearrangement of alkyl substituents bound to the arene.

to the preparation of half-sandwich arene complexes and metallocenes may sometimes be used in place of metal halides (C_5H_5 abstraction) to provide mixed arene–cyclopentadienyl sandwiches. In the interim, many more synthetic routes have emerged for the preparation of arene complexes of transition metals (and some main group elements). Amongst these, metal-atom vapour/ligand co-condensation (Figure 7.35) deserves special comment because it is in arene chemistry that it finds its widest application.

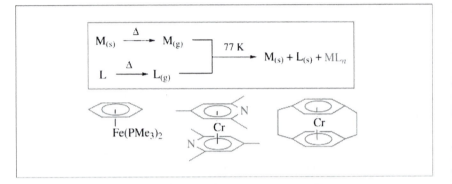

Figure 7.35 Metal atom–ligand co-condensation

The metal and ligand are both vaporized and then co-condensed together, usually at liquid nitrogen temperature (77 K). In addition to finely divided (and pyrophoric) bulk metal, organometallic complexes may generally be extracted from the resulting condensate. Although this non-routine method requires sophisticated equipment, the strength lies in accessing compounds not available by conventional 'wet' routes, including those shown in Figure 7.35. The high energy required to vaporize some metals presently excludes them from this approach.

Cyclohexadienes are readily available *via* Birch reduction of arenes, and may also serve as precursors *via* metal-mediated dehydrogenation, *e.g.* the majority of half-sandwich arene chemistry based on ruthenium arises from the binuclear complexes $Ru_2(\mu\text{-Cl})_2Cl_2(\eta^6\text{-arene})_2$ (arene = $MeC_6H_4Pr^i$, C_6H_6) which result from the reaction of $RuCl_3.xH_2O$ with α-phellandrene or cyclohexadiene (Figure 7.36). The cymene or benzene ligands may be replaced by more substituted arenes (*e.g.* $C_6H_nMe_{6-n}$, n = 3, 6) under thermally forcing conditions. These are also photolabile, whilst the chloride bridges are easily cleaved by nucleophilic ligands to provide mononuclear half-sandwich derivatives.

The bonding in bis(arene) complexes in terms of orbital symmetry is essentially analogous to that of metallocenes: the frontier orbitals of benzene, familiar from any organic textbook, are of correct symmetry (σ, π and δ) for interaction with all the s, p and d orbitals of a metal centre. In energetic terms, however, there is a somewhat closer match in energy between the Cr^0/C_6H_6 combination than $Fe^{2+}/C_5H_5^-$, leading to

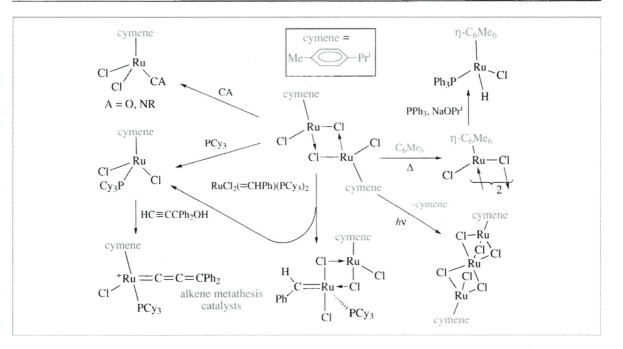

Figure 7.36 Synthesis and reactions of a ruthenium half-sandwich arene complex

stronger δ-bonding (e_{2g} + d_{xy}, $d_{x^2-y^2}$). Figure 7.37 shows the molecular orbital scheme for dibenzenechromium, from which the clear HOMO–LUMO gap for the d^6 18VE species helps to rationalize the observed stability of $M(C_6R_6)_2$ (M = Cr, Mo, W, Fe^{2+}, Ru^{2+}, Os^{2+}). Note the topological similarity of the central part of this scheme to Figure 7.23, although it must be emphasized that the exact energetics and orbital contributions will vary with the nature of the metal and the carbocyclic ring.

Whilst metallocenes have an enormously rich chemistry, in the case of arene coordination it is the half-sandwich ('piano-stool') class of complex which finds the widest utility in organic synthesis, in particular within the chemistry of $Cr(CO)_3(\eta^6$-arene) complexes. These are generally prepared directly from the thermal reaction of $Cr(CO)_6$ and arene in a high-boiling solvent (*e.g.* Bu_2O, 2-picoline). For less robust arenes, the labile complexes $Cr(L)_3(CO)_3$ (L = NH_3, THF, MeCN, py) may be employed under milder conditions (Figure 7.38).

As with cyclopentadienyl metal carbonyls, the coordination of the net donor arene ligand *trans* to three π-acidic CO ligands confers considerable thermodynamic stability. Relative to the free arene, the coordination of a '$Cr(CO)_3$' unit to an arene face removes electron density, to an extent comparable to the inclusion of a nitro ring substituent. The chemical implications of this electron drift from the arene to chromium, and the basis for much chemistry, are an increase in the acidity of protons bound to the arene ring or to the α- and β-carbons of substituents, and

Figure 7.37 (right) Molecular orbital description of dibenzenechromium, $Cr(\eta\text{-}C_6H_6)_2$

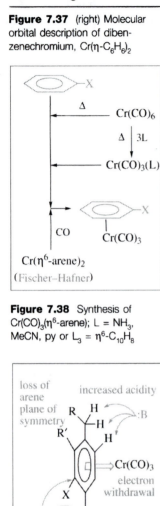

Figure 7.38 Synthesis of $Cr(CO)_3(\eta^6\text{-arene})$; L = NH_3, MeCN, py or $L_3 = \eta^6\text{-}C_{10}H_8$

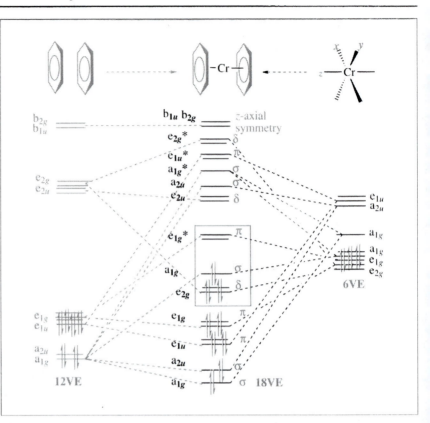

Figure 7.39 Chemical implications of η^6-arene coordination

an activation towards attack by nucleophiles *exo* to the metal (Figure 7.39).

The nucleofugacity of leaving groups bound to either the α- or β-carbons is enhanced by the metal centre being able to stabilize the resulting carbocation. However, it is the increase in reactivity towards nucleophilic substitution at the ring itself which finds widest application since S_NAr reactions for free arenes are typically difficult (however, see Figure 4.10). These various activations are useful in their own right, but a further dimension is added if the free arene has no element of symmetry other than the C_6 plane. In such arenes (*e.g.* 1,2- or 1,3-XYC_6H_4), the faces are diastereotopic, *i.e.* association of a reagent (in this case metal) to one or other face will result in enantiomers. If either the metal centre or an arene substitutent is chiral, then (possibly separable) diastereomers will result (see coordination of prochiral alkenes, Figure 5.6). Similar arguments apply to the benzylic protons adjacent to coordinated prochiral arenes, which may show diastereoselective reactivity.

Figure 7.40 illustrates transformations which may be mediated by coordination of an arene to chromium. As with cyclopentadienyl ligands, it is possible to metallate the ring with BuLi (often regioselectively) to provide η^6-aryllithium reagents which may be treated with a range

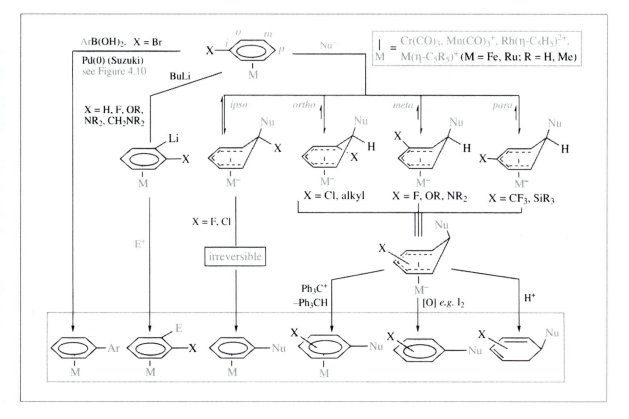

of electrophiles (*cf.* Figure 7.30), *e.g.* with ClPPh$_2$ a *hexahapto* complex of PPh$_3$ (E = PPh$_2$) is obtained. Chromium-complexed aryl halides may also be employed in the palladium-catalysed coupling with arylboronic acids (Figure 4.10). The most extensively studied class of reactions involves direct attack by nucleophiles at the *exo* face of the arene, which results in the formation of an anionic cyclohexadienyl complex (*cf.* [Cr(CO)$_3$Cp]$^-$). In general, this addition is reversible and, in the case of monosubstituted arenes, attack is possible at all four sites (*ipso, ortho, meta, para*); however, for haloarenes, *ispo* attack is usually followed irreversibly by loss of halide to effect S$_N$Ar. More generally, however, the product of nucleophilic attack may be trapped either by protonation to provide the substituted cyclohexadiene, or hydride abstraction (Ph$_3$C$^+$) to provide a new disubstituted arene complex. Because the nucleophilic attack is reversible, the majority of other electrophiles simply attack the uncoordinated nucleophile to regenerate the starting complex. A special case involves sulfur-stabilized anions, in which the addition is irreversible and subsequent quenching by electrophiles proceeds *via* attack at the metal followed by migration to the ring (reductive elimination). In the case of alkyl halides, initial migration to a carbonyl first followed by reductive elimination provides acyl cyclohexadienes (Figure 7.41).

Figure 7.40 Metal-mediated arene functionalization *via* *hexahapto* coordination

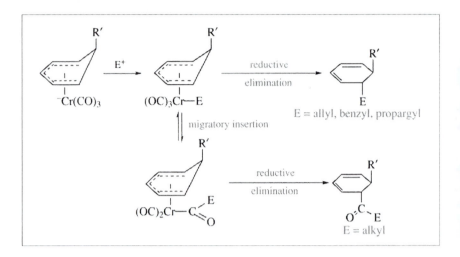

Figure 7.41 Alkylation of cyclohexadienyl chromium complexes; R' = CH$_x$(SR)$_{3-x}$

The toxicity of chromium salts and carbonyls confers some urgency upon the search for catalytic protocols or stoichiometric processes based on less toxic metals.

In addition to the 'Cr(CO)$_3$,' group, a range of other (primarily d^6 ML$_3$) fragments have been employed for the coordinative activation of arenes (Figure 7.42). *En retarde* of alkene chemistry, catalytic protocols for the modification of arenes *via hexahapto* coordination are at present less well evolved; however, this will no doubt follow. Primarily this may be traced to the energetics of introducing and liberating the arene from the metal centre.

7.7.1 Reduced Hapticity

Arenes have a total of 6VE with which to provide a metal centre; however, if this were to contravene the 18-electron rule, then the possibility of reduced hapticity arises. Thus two-electron reduction of the 18VE cation [Ru(η-C$_6$Me$_6$)$_2$]$^{2+}$ provides the neutral ruthenium(0) complex Ru(η^4-C$_6$Me$_6$)(η-C$_6$Me$_6$) wherein one arene coordinates as a buta-1,3-diene. The loss of aromatic stabilization is less drastic for naphthalenes (10π), which on 'ring slippage' retain the 6π system of the uncomplexed ring (*cf.* indenyl effect). Thus [Ru(η-C$_6$H$_6$)(C$_{10}$Me$_8$)]$^{2+}$ upon reduction gives a complex where it is the naphthalene ligand which exclusively adopts the *tetrahapto* coordination, such that the intact uncoordinated benzo group may then be coordinated to chromium (Figure 7.42). Because of this ring slippage lability of naphthalenes (and anthracenes), η^6-naphthalene complexes are widely used as isolable precursors for subsequent arene exchange reactions, *e.g.* Cr(CO)$_3$(η^6-C$_{10}$H$_8$) (Figure 7.38).

The previous discussion focused on coordination of arenes to metal centres ligated by π-acids, the net result being that the role of the arene was primarily as a donor and thereby activated towards nucleophilic attack. A complementary situation arises when the metal centre has only pure σ-donor co-ligands. The resulting retrodative capacity of the metal

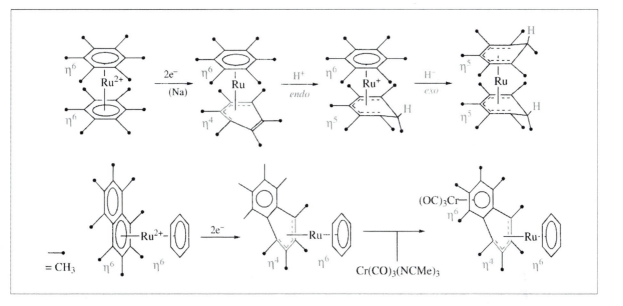

Figure 7.42 *Tetrahapto* arene coordination

centre is then best alleviated by interaction with a π-acceptor ligand. In the case of arenes, this is achieved, quite remarkably, by coordination of only one C=C double bond, at the expense of aromaticity. The ability to (temporarily) induce alkenic reactivity in an aromatic compound is synthetically valuable, and this is most mature in the case of the '$Os(NH_3)_5^{2+}$' fragment (Figure 7.43), which binds strongly to π-acceptor ligands (relativistic enhancement, Chapter 1) but essentially ignores functional groups which can only serve as σ-donors (avoiding the need for protection/deprotection protocols for esters, amides, ethers, alcohols, water and amines). Thus reduction of the trivalent precursor (for $[Os(NH_3)_6]^{3+/2+}$, $E^\circ = -0.78$ V *vs.* NHE) in propanone (acetone) provides a propanone complex, π-bound through the C=O bond rather than conventional O-coordination (σ-donor). In this case, the *dihapto* co-ordinated arene is activated towards electrophilic substrates (attack *anti* to the metal), including even weak electrophiles such as Michael acceptors ($H_2C=CHR$; R = CN, CO_2Me, COMe). Alternatively, the perturbation of the π-system induces 1,3-diene character on the four unbound carbons, *e.g.* deuteration (D_2, Pd/C) of the benzene complex provides the d^4 cyclohexene complex (all D *anti* to Os). The modified arene is then liberated by oxidation to Os^{III}, which does not bind π-acids as strongly (*e.g.* the η^2-Me_2NPh and 2,6-dimethylpyridine complexes rearrange, on one-electron oxidation, to σ-amine complexes). Conventional alkenes, however, *e.g.* styrene, will bind in preference to the C=C bond. With six C–C bonds to choose from, coordination normally occurs so as to minimize steric pressures and the disruption of the π-system, although migration around the ring may occur (typically $\Delta G^{\ddagger} = 50$–75 kJ mol^{-1}).

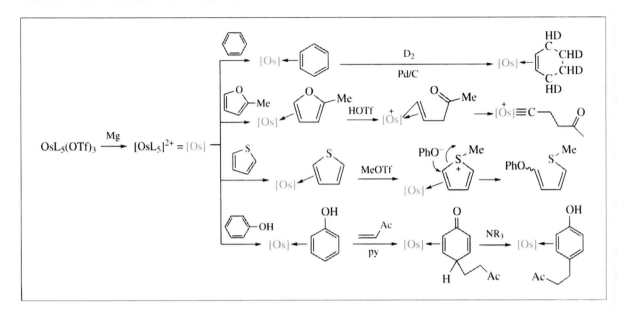

Figure 7.43 *Dihapto* arene coordination: synthetic implications; L = NH$_3$; Tf = SO$_2$CF$_3$

(a) Os$_3$(CO)$_9$(μ_3-C$_6$H$_6$)

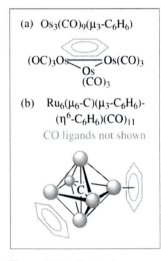

(OC)$_3$Os———Os(CO)$_3$
        Os
       (CO)$_3$

(b) Ru$_6$(μ_6-C)(μ_3-C$_6$H$_6$)-(η^6-C$_6$H$_6$)(CO)$_{11}$

CO ligands not shown

Figure 7.44 Cluster face-capping (μ_3:η^2,η^2,η^2) arenes

The cluster–surface analogy (Figure 2.19) also extends in principle to arenes (and other η^n-C$_n$R$_n$ ligands), including the osmium and ruthenium clusters depicted in Figure 7.44. The hexaruthenium example contains one *hexahapto* benzene ligated to one ruthenium, whilst the second bridges a triruthenium face (μ_3:η^2,η^2,η^2). Facile rotation of the arene (NMR), however, argues against a localized cyclohexatriene bonding description.

7.8   Cycloheptatrienyls, η^7-C$_7$R$_7$

The tropylium cation is a stable 6π aromatic system, allowing salts to be prepared with non-coordinating anions. For Groups 4–6, *heptahapto* planar coordination is commonly encountered including examples with <18VE. However, for later transition metals (greater d occupancies), binding modes of reduced hapticity are common (as with cyclooctatetraene, Figure 6.31). By far the majority of η^7-C$_7$H$_7$ complexes originate from readily available cycloheptatriene, *via hexahapto* complexes of this neutral ligand from which hydrogen (generally as H$^-$, but occasionally as H• or H$^+$) is abstracted. This is illustrated by the synthesis of [Mo(CO)$_3$(η-C$_7$H$_7$)]$^+$ (Figure 7.45).

By microscopic reversibility, the possibility of nucleophilic attack at the ring arises, in particular for cationic complexes, and this generally occurs at an *exo* site, allowing the synthesis of functionalized cyclohexatriene complexes. When it arises, the reduced hapticity of the C$_7$H$_7$ ring *to one metal* may allow the introduction of a second metal such that the C$_7$H$_7$ ring serves as a bridging ligand (Figures 7.45, 7.46). Furthermore,

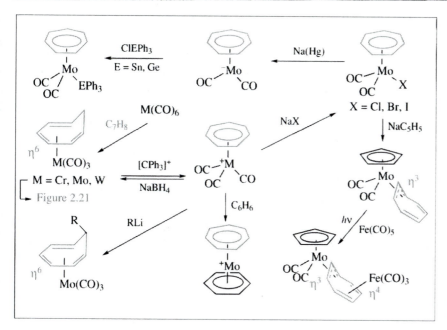

Figure 7.45 Synthesis and reactivity of $[Mo(CO)_3(\eta\text{-}C_7H_7)]^+$

even for *heptahapto* coordination, there are indications that co-ligand substitution processes may proceed *via pentahapto* coordinated intermediates, allowing an associative mechanism to operate (*cf.* indenyl effect).

Figure 7.46 Various coordination modes for C_7H_7 and C_7H_8 ligands

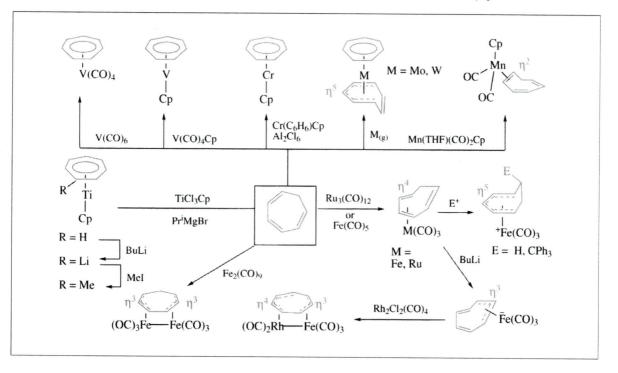

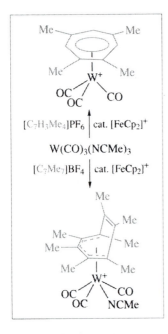

Figure 7.47 Steric effects in tropylium coordination

Whilst the chemistry of $\eta^5\text{-}C_5Me_5$ and $\eta^6\text{-}C_6Me_6$ ligands is enormously rich, a problem arises with the C_7Me_7 ligand: as n increases for $\eta^n\text{-}C_nR_n$ ligands, so does the steric repulsion between the adjacent R groups. Furthermore, the increase in ring size requires that these substituents bend towards the metal, increasing intersubstituent repulsion. Thus, to date, no $\eta^7\text{-}C_7Me_7$ complexes have been isolated, there being a tendency towards reduced hapticity (*i.e.* $\eta^5\text{-}C_7Me_7$) whereby the pseudo-aromaticity is sacrificed in favour of reducing steric repulsions (Figure 7.47). Indeed, even the free cation $[C_7Me_7]^+$ adopts a non-planar (boat) structure.

Azulene (Figure 7.48) is an intriguing variant wherein the cycloheptatrienyl unit is fused with a five-membered ring. The organometallic chemistry of this hydrocarbon and related derivatives is less intensely studied but includes the 18VE derivatives shown, obtained from reactions with simple binary carbonyls.

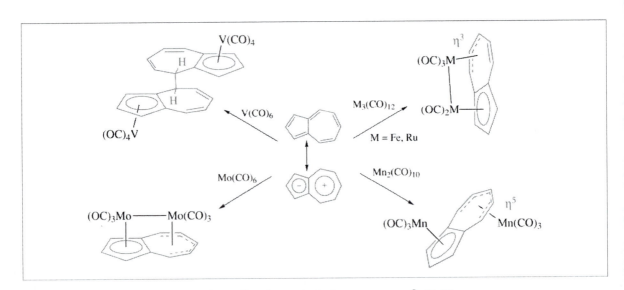

Figure 7.48 Reactions of azulene with binary carbonyls

7.9   Cyclooctatetraenes, $\eta^8\text{-}C_8R_8$

Cyclooctatetraene (cot) behaves something like an octopus when interacting with transition metals in that a wide range of hapticities (η^2, η^4, η^6 and η^8, Figure 7.49) are commonly observed or (cautiously!) inferred from spectroscopic data. On moving from Group 4 to Group 10 there is a progressive trend from octahapticity to dihapticity, as the central metal itself provides increasingly more valence electrons and therefore requires less from the C_8R_8 ligand. Since the 18-electron rule is redundant for lanthanides and actinides, *octahapto* coordination is most common for these elements, including the first C_8H_8 sandwich 'uranocene', $U(\eta$-$C_8H_8)_2$.

The first complex of cyclooctatetraene prepared by Stone in fact involved static *tetrahapto* (buta-1,3-diene-like) C_8H_8 coordination in the solid state, but dynamic migration of the iron centre around the ring in solution (a single ^1H NMR resonance). This dichotomy presaged the fluxionality encountered in many subsequent complexes of cyclooctatetraenes, wherein the hapticity does not follow unambiguously from solution spectroscopic data, *e.g.* M(cot)$_2$ (M = Fe, Ru; Figure 7.50) also have single ^1H NMR resonances at room temperature, despite two distinct coordination modes (η^6, η^4) for the rings in the solid state.

The cyclooctatetraene ligand is generally introduced in one of two ways: either as the neutral hydrocarbon or alternatively *via* initial reduction to the aromatic (10π) dianion by treatment with a Group 1 metal (Figure 7.51).

7.10 Heteroarenes ('Inorganometallic' Chemistry)

We have only considered *polyhapto* carbocyclic ligands; however, the field of aromatic heterocycles is immense and conceptually simplified by recognizing simple isoelectronic relationships between 'CH' and various (possibly substituent bearing or charged) heteroatoms. Many such

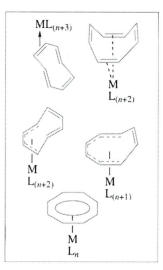

Figure 7.49 Variable hapticity in cyclooctatetraene coordination

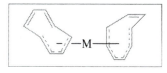

Figure 7.50 M(η^4-C_8H_8)(η^6-C_8H_8) (M = Fe, Ru)

Figure 7.51 (Below) Synthesis of cyclooctatetraene complexes

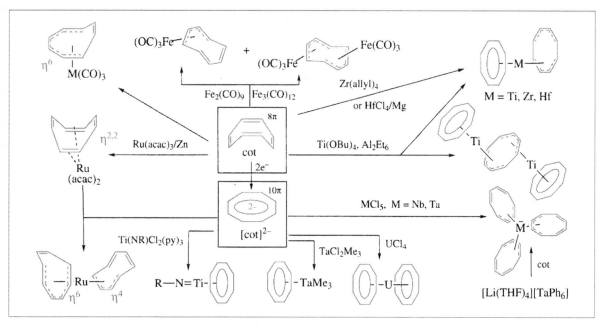

heteroaromatics can indeed mimic the type of *polyhapto* coordination displayed by the η^n-C_nR_n ligands. Such molecules fall outside the scope of this introductory text; however, some of the elegant examples illustrated in Figure 7.52 should whet the appetite for further reading. Taking the analogy to the extreme, we have the entirely 'inorganic' examples of borazine ($B_3N_3R_6$), which is isolable as a free molecule, and *cyclo*-P_6, which is only stable when coordinated to a metal centre.

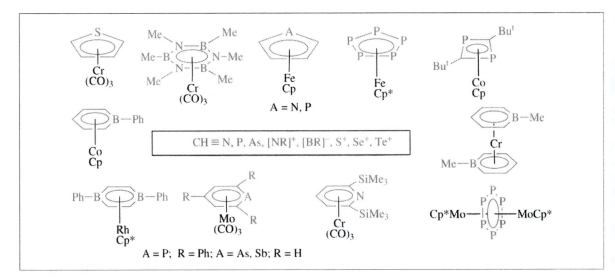

Figure 7.52 Inorganometallic analogues of η^n-C_nH_n ligands

Abbreviations

acac   propane-2,4-dionate, acetylacetonate
BDE   bond dissociation enthalpy
cod   cyclooctadiene
Cp η-C_5H_5, *pentahapto* cyclopentadienyl
Cp* η-C_5Me_5, *pentahapto* pentamethylcyclopentadienyl
DMF   *N,N*-dimethylformamide
dmpe   1,2-bis(dimethylphosphino)ethane
dppe   1,2-bis(diphenylphosphino)ethane
EAN   effective atomic number
en   1,2-diaminoethane, ethylenediamine
HMPA   hexamethylphosphoric triamide, $O=P(NMe_2)_3$
HOMO   highest occupied molecular orbital
HSAB   hard and soft acid and base
IR   infrared
LUMO   lowest unoccupied molecular orbital
ML_n a general metal 'M' ligated by n ligands
nbd   norbornadiene
NMR   nuclear magnetic resonance
nVE n valence electrons
PE   Pauling electronegativity
R   a general organic group (Me = methyl; Et = ethyl; Pr = propyl;
            Bu = butyl; Cy = cyclohexyl; Ar = aryl; Ph = phenyl; Bz = benzyl;
            pz = pyrazol-1-yl; Ad = adenosyl)
Tf   trifluoromethanesulfonyl (OTf = trifluoromethanesulfonate)
THF   tetrahydrofuran
tmeda   *N,N,N',N'*-tetramethyl-1,2-diaminoethane,
            N,N,N',N'-tetramethylethylenediamine
TPP   tetraphenylporphyrinate

$\Delta_{o/t}$ crystal field splitting parameter for octahedral/tetrahedral field
η^n *n-hapto* ligand
μ_n a ligand bridging n metals

Subject Index

CPSIA information can be obtained at www.ICGtesting.com
Printed in the USA
LVOW09s0825161214

419035LV00024B/691/P